D.P. LANE

D.P. LANE

The Role of Apoptosis in Development, Tissue Homeostasis and Malignancy

Death from inside out

The Role of Apoptosis
in Development, Tissue Homeostasis
and Malignancy

Death from inside out

EDITED BY

T.M. Dexter
CRC Department of Experimental Haematology
Paterson Institute for Cancer Research
Manchester, UK

M.C. Raff
MRC Laboratory for Molecular Cell Biology
University College London
London, UK

and

A.H. Wyllie
CRC Laboratories
University Medical School
Edinburgh, UK

 The Royal Society

 CHAPMAN & HALL
London · Glasgow · Weinheim · New York · Tokyo · Melbourne · Madras

Published by Chapman & Hall, 2–6 Boundary Row, London SE1 8HN, UK

Chapman & Hall, 2–6 Boundary Row, London SE1 8HN, UK

Blackie Academic & Professional, Wester Cleddens Road, Bishopbriggs, Glasgow G64 2NZ, UK

Chapman & Hall GmbH, Pappelallee 3, 69469 Weinheim, Germany

Chapman & Hall USA, 115 Fifth Avenue, New York, NY 10003, USA

Chapman & Hall Japan, ITP-Japan, Kyowa Building, 3F, 2-2-1 Hirakawacho, Chiyoda-ku, Tokyo 102, Japan

Chapman & Hall Australia, 102 Dodds Street, South Melbourne, Victoria 3205, Australia

Chapman & Hall India, R. Seshadri, 32 Second Main Road, CIT East, Madras 600 035, India

© 1995 The Royal Society

Printed in Great Britain by TJ Press, Padstow,

ISBN 0 412 63810 X

A catalogue record for this book is available from the British Library

Printed on acid-free text paper, manufactured in accordance with ANSI/NISO Z39.48-1992 (Permanence of Paper).

Contents

CLINICAL APPLICATIONS

Contributors

J.M. Abrams   Howard Hughes Medical Institute, Department of Brain and Cognitive Sciences and Department of Biology, Massachusetts Institute of Technology, Cambridge, Massachusetts 02139, USA

Bruno Amati   Biochemistry of the Cell Nucleus Laboratory, Imperial Cancer Research Fund, PO Box 123, 44 Lincoln's Inn Fields, London, WC2A 3PX

B.A. Barres   MRC Developmental Neurobiology Programme, MRC Laboratory for Molecular Cell Biology and Biology Department, University College London, London, WC1E 6BT; *Current address*: Department of Neurobiology, Stanford University School of Medicine, Stanford, California, 94305–5401, USA

Martin Bennett   Biochemistry of the Cell Nucleus Laboratory, Imperial Cancer Research Fund, PO Box 123, 44 Lincoln's Inn Fields, London, WC2A 3PX

J.F. Burne   MRC Developmental Neurobiology Programme, MRC Laboratory for Molecular Cell Biology and Biology Department, University College London, London, WC1E 6BT

H.S.R. Coles   MRC Developmental Neurobiology Programme, MRC Laboratory for Molecular Cell Biology and Biology Department, University College London, London, WC1E 6BT

Suzanne Cory   The Walter and Eliza Hall Institute of Medical Research, PO Box Royal Melbourne Hospital, Victoria 3050, Australia

G.J. Cowling   CRC Department of Experimental Haematology, Paterson Institute for Cancer Research, Christie Hospital (NHS) Trust, Manchester, M20 9BX

T.M. Dexter   CRC Department of Experimental Haematology, Paterson Institute for Cancer Research, Christie Hospital (NHS) Trust, Manchester, M20 9BX

I. Dransfield   Respiratory Medicine Unit, Department of Medicine (RIE), University of Edinburgh Royal Infirmary, Lauriston Place, Edinburgh, EH3 9YW

Gerard Evan   Biochemistry of the Cell Nucleus Laboratory, Imperial Cancer Research Fund, PO Box 123, 44 Lincoln's Inn Fields, London, WC2A 3PX

Abdallah Fanidi   Biochemistry of the Cell Nucleus Laboratory, Imperial Cancer Research Fund, PO Box 123, 44 Lincoln's Inn Fields, London, WC2A 3PX

T.C. Fisher   Cancer Research Campaign Molecular and Cellular Pharmacology Group, The School of Biological Sciences, University of Manchester, Stopford Building (G38), Manchester M13 9PT

James L. Franklin   Department of Molecular Biology and Pharmacology, Washington University School of Medicine, 660 South Euclid Avenue, St Louis, Missouri 63110, USA

P. Golstein   Centre d'Immunologie INSERM-CNRS de Marseille-Luminy, Case 906, 13288 Marseille Cedex 9, France

M.E. Grether   Howard Hughes Medical Institute, Department of Brain and Cognitive Sciences and Department of Biology, Massachusetts Institute of Technology, Cambridge, Massachusetts 02139, USA

P.A. Hall   Department of Pathology, University of Dundee Medical School, Dundee DD1 9SY

Elizabeth Harrington	Biochemistry of the Cell Nucleus Laboratory, Imperial Cancer Research Fund, PO Box 123, 44 Lincoln's Inn Fields, London, WC2A 3PX
Alan W. Harris	The Walter and Eliza Hall Institute of Medical Research, PO Box Royal Melbourne Hospital, Victoria 3050, Australia
C. Haslett	Respiratory Medicine Unit, Department of Medicine (RIE), University of Edinburgh Royal Infirmary, Lauriston Place, Edinburgh, EH3 9YW
Michael O. Hengartner	Howard Hughes Medical Institute, Department of Biology, Room 56–629, Massachusetts Institute of Technology, 77 Massachusetts Avenue, Cambridge, Massachusetts 02139, USA
J.A. Hickman	Cancer Research Campaign Molecular and Cellular Pharmacology Group, The School of Biological Sciences, University of Manchester, Stopford Building (G38), Manchester M13 9PT
Tasuku Honjo	Department of Medical Chemistry, Faculty of Medicine, Kyoto University, Kyoto 606, Japan
H. Robert Horvitz	Howard Hughes Medical Institute, Department of Biology, Room 56–629, Massachusetts Institute of Technology, 77 Massachusetts Avenue, Cambridge, Massachusetts 02139, USA
Ted Hupp	Cancer Research Campaign Laboratories, University of Dundee, Dundee DD1 4HN
Y. Ishizaki	MRC Developmental Neurobiology Programme, MRC Laboratory for Molecular Cell Biology and Biology Department, University College London, London, WC1E 6BT
M.D. Jacobson	MRC Developmental Neurobiology Programme, MRC Laboratory for Molecular Cell Biology and Biology Department, University College London, London, WC1E 6BT
Eugene M. Johnson Jr	Department of Molecular Biology and Pharmacology, Washington, University School of Medicine, 660 South Euclid Avenue, St Louis, Missouri 63110, USA
Andrew Kung	Department of Biological Sciences, Stanford University, Stanford, California 94305, USA
Hartmut Land	Biochemistry of the Cell Nucleus Laboratory, Imperial Cancer Research Fund, PO Box 123, 44 Lincoln's Inn Fields, London, WC2A 3PX
D.P. Lane	Cancer Research Campaign Laboratories, University of Dundee, Dundee, DD1 4HN
Xin Lu	Ludwig Institute for Cancer Research, St Mary's Hospital Medical School, London W2 3PG
M.-F. Luciani	Centre d'Immunologie INSERM-CNRS de Marseille-Luminy, Case 906, 13288 Marseille Cedex 9, France
A.J. Merritt	Cancer Research Campaign Molecular and Cellular Pharmacology Group, The School of Biological Sciences, University of Manchester, Stopford Building (G38), Manchester M13 9PT *and* CRC Department of Epithelial Biology, Paterson Institute for Cancer Research, Christie Hospital (NHS) Trust, Wilmslow Road, Manchester, M20 9BX
L.C. Meagher	Respiratory Medicine Unit, Department of Medicine (RIE), University of Edinburgh Royal Infirmary, Lauriston Place, Edinburgh, EH3 9YW
Masao Murakami	Department of Medical Chemistry, Faculty of Medicine, Kyoto University, Kyoto 606, Japan
Shigekazu Nagata	Osaka Bioscience Institute, 6–2–4 Furuedai, Suita, Osaka 565, Japan
Sazuku Nisitani	Department of Medical Chemistry, Faculty of Medicine, Kyoto University, Kyoto 606, Japan

C.S. Potten CRC Department of Epithelial Biology, Paterson Institute for Cancer Research, Christie Hospital (NHS) Trust, Wilmslow Road, Manchester, M20 9BX

M.C. Raff MRC Developmental Neurobiology Programme, MRC Laboratory for Molecular Cell Biology and Biology Department, University College London, London, WC1E 6BT

J.S. Savill Division of Renal & Inflammatory Disease, Department of Medicine, University Hospital, Nottingham, NG7 2UH

Robert T. Schimke Department of Biological Sciences, Stanford University, Stanford, California 94305, USA

Rakesh Sharma Department of Biological Sciences, Stanford University, Stanford, California 94305, USA

Steven S. Sherwood Department of Biological Sciences, Stanford University, Stanford, California 94305, USA

Jamie Sheridan Department of Biological Sciences, Stanford University, Stanford, California 94305, USA

H. Steller Howard Hughes Medical Institute, Department of Brain and Cognitive Sciences and Department of Biology, Massachusetts Institute of Technology, Cambridge, Massachusetts 02139, USA

M. Stern Department of Respiratory Medicine, Royal Postgraduate Medical School, Hammersmith Hospital, Du Cane Road, London, W12 0NN

Andrew Strasser The Walter and Eliza Hall Institute of Medical Research, PO Box Royal Melbourne Hospital, Victoria 3050, Australia

Takeshi Tsubata Department of Medical Chemistry, Faculty of Medicine, Kyoto University, Kyoto 606, Japan

K. White Howard Hughes Medical Institute, Department of Brain and Cognitive Sciences and Department of Biology, Massachusetts Institute of Technology, Cambridge, Massachusetts 02139, USA

M.K.B. Whyte Department of Respiratory Medicine, Royal Postgraduate Medical School, Hammersmith Hospital, Du Cane Road, London, W12 0NN

Andrew H. Wyllie Cancer Research Campaign Laboratories, Department of Pathology, University Medical School, Edinburgh, EH8 9AG

Preface

The past five years have witnessed a remarkable development of interest in cell death 'from inside out'. After 30 years of relative obscurity, its quantitative importance in the building and maintenance of normal tissues, the subtle strategies involved in its regulation, and its significance in the pathogenesis of diseases of major social importance are becoming clear. Moreover, because a distinct set of biological events is involved in this death, these events themselves become reasonable targets for new pharmacological agents in the treatment of cancer. The articles in this volume summarize the contents of a discussion meeting held at the Royal Society on 23 and 24 February 1994. The authors are a distinguished international group from a variety of disciplines in biology and medicine and hopefully their articles will convey something of the excitement of this fast-moving field. The three organizers are enormously indebted to all the contributors for the enthusiasm with which they delivered their talks, shared in discussion, and finally committed their contributions to these printed pages. We would also like to acknowledge the gracious way in which the Royal Society hosted the meeting, and in particular Mary Manning for making it the trouble-free and enjoyable experience that it was, and Janet Clifford and Simon Gribbin for skillfully managing the editorial processing of this volume.

June 1994

Michael Dexter

Martin Raff

Andrew Wyllie

1

Death from inside out: an overview

ANDREW H. WYLLIE

Cancer Research Campaign Laboratories, Department of Pathology, University Medical School, Edinburgh EH8 9AG, U.K.

SUMMARY

Although a type of cell death strategically suited to participating in developmental processes has been well known for nearly thirty years, it is only in the recent past that the extraordinary ubiquity of such death has been appreciated. Apoptosis, a term first employed to describe such death defined in structural terms, is associated with a stereotyped set of effector processes, and is driven by genes most of which are familiar as oncogenes or oncosuppressor genes. Dysregulation of apoptosis leads to diseases of enormous social importance such as cancer and AIDS.

1. INTRODUCTION

The papers that follow in this volume deal with various aspects of cell death 'from inside out'. By this, their authors mean that the dying cells they study are contributing in an active way to the triggering and execution of the processes that lead to their own demise. This view of cell death, prevalent among developmental biologists for many decades (Saunders 1966), has taken some time to gain credibility in other branches of biology and pathology. Perhaps this was in part because these disciplines were already familiar with a well-described process (necrosis) characterized by breakdown of cellular energy supply, failure of cellular volume homeostasis, plasma membrane rupture and an ensuing acute inflammatory reaction (Trump & Beresky 1994): events that did not seem congruent with the subtle removal of cells during the sculpting of developing tissues. Perhaps also, even in developmental biology, the cell biological tools were for long not available with which to dissect the mechanisms of death. Within the past four or five years the situation has changed in a scale and at a rate that few could have predicted. A set of genes has been identified that appear to activate or modify a stereotyped programme of effector events, orchestrated to ensure both the rapid demise and practically instantaneous recognition and removal of the dying cell. Moreover, these genes are members of families that are highly conserved between species, suggesting that this process of death may be fundamental to the management of cells and tissues in metazoan organisms. Whereas some of these genes were hitherto unknown, others are extremely familiar in the context of cell proliferation and cancer. A picture is emerging, although not yet completely formed, that death 'from inside out' is a process not too dissimilar in organization to cell division, perhaps using analogous or even identical signalling and effector molecules.

2. MORPHOLOGICAL CONSIDERATIONS

The structural aspects of 'death from inside out' were the first to be resolved, and have now been reviewed many times (Wyllie *et al.* 1980; Arends & Wyllie 1991). Key elements are shrinkage of cell volume, loss of specialized plasma membrane regions such as microvilli, morphological conservation of most cytoplasmic organelles, progressive perinuclear chromatin condensation, and exposure of surface signals that can facilitate engulfment of the dying cell by adjacent phagocytes. This sequence of events was first recognized clearly in hepatocytes *in vivo* (Kerr 1971) and the process underlying them was named apoptosis in 1972, in recognition of its wide significance in tissue homeostasis (Kerr *et al.* 1972).

A corollary of the organization of apoptosis is its inconspicuous nature in tissue sections, even when it is responsible for extensive and rapid cell loss. The time interval between commitment to death and the appearance of the first characteristic cellular features varies according to cell type and lethal stimulus, but there is agreement that the time from first appearance of the structural changes until the dying cell disappears within the phagosome of the ingesting cell, may be a matter of an hour or two, perhaps less. New methods for identification of the changes in the nuclei of apoptotic cells may render them more conspicuous (Gavrieli *et al.* 1992; Ansari *et al.* 1993), but the accurate assessment of the extent of cell death in tissues usually requires scanning of several thousand normal cells, as the proportion of apoptotic cells is often much less than 1%. Because of the short 'washout time' for which apoptotic cells are recognizable, and their total disappearance thereafter, such low percentages can nonetheless be responsible for major reductions in total cell number in a tissue (Howie *et al.* 1994).

3. EFFECTOR MECHANISMS

Chromatin condensation is associated with evidence of chromatin cleavage, first to fragments corresponding in size (50 and 300 kb) to the loop and rosette domains into which chromatin is organized (Roy *et al.* 1992). Thereafter, many cells show cleavage down to mononucleosome and oligonucleosome size, to engender the familiar 'ladder' pattern on DNA electrophoresis (Wyllie 1980). Attempts to purify the endonucleases concerned have identified a neutral, Mg-dependent nuclease of around 19 kDa that bears immunological similarity if not identity to DNAse I (Gaido & Cidlowski 1991; Peitsch *et al.* 1993). Conversely, in other cell types, this nuclease is not present in measurable quantity, but there is an acid nuclease of *ca.* 35 kDa, that can cleave DNA independent of ambient calcium or magnesium concentration (Barry & Eastman 1993). Several other proteins appear in apoptotic cells, but are less evident or absent in their viable counterparts. It is tempting to assume that they are effector proteins of apoptosis, particularly when their functions seem to be what is necessary to bring about the cellular changes known to occur in apoptosis. At present, however, there are few defined molecular species whose roles as effectors of apoptosis have been stringently established. Thus, apoptotic cells appear to contain a site-specific ribonuclease that cleaves 28*S* rRNA in a non-processive manner (Houge *et al.* 1993). Apoptotic bodies contain proteins that are insoluble in ionic detergent, apparently as a result of cross-linking by a tissue-type transglutaminase (Fesus *et al.* 1991). There are a variety of membrane changes that may permit expression of the signal for phagocytosis (Duvall *et al.* 1985; Savill *et al.* 1990; Fadok *et al.* 1992). Some apoptotic cells express the sulphoprotease clusterin (TRPM-2) on the plasma membrane (Monpetit *et al.* 1986), but this is unlikely to be either necessary or specific for apoptosis. Many cells do not enter apoptosis if RNA and protein synthesis are blocked, but this also is not universal. Protease inhibitors block apoptosis in several cell types (Gorczyca *et al.* 1992), and it is interesting that specific proteases are essential both for the programmed death of cells in the nematode *Caenorhabditis elegans* (Yuan *et al.* 1993) and for the death of the targets of killing by cytotoxic T cells (Shi *et al.* 1992). The lethal agents released from the granules of these cells have long been suspected of including a cocktail of effector molecules required for apoptosis. New, currently anonymous molecular species associated temporally with apoptosis are being intensively searched for by subtractive and differential display techniques.

Despite the fact that many of these effector molecules are incompletely defined, it is already clear that cells differ greatly in their content of some of them. We observed that fibroblast lines contained widely differing activities of endogenous nuclear endonuclease, even although these lines all derived from a common parental stock (Arends *et al.* 1993). Moreover, the nuclear endonuclease activity correlated broadly with the ease with which cells from the different lines underwent apoptosis during log-phase growth in culture. Thus it is possible that cells exist in at least two different states with regard to apoptosis, one in which they are endowed with the necessary effector proteins (we have called this the *primed* state), and would enter the process if suitably *triggered*, and another in which new effectors would need to be synthesized before apoptosis could proceed (the *unprimed* state) (Arends & Wyllie 1991). In these experiments, the various related fibroblast lines were generated from their common parent by independent transfection of a variety of oncogenes. Because unwanted and undetected events might have occurred during the selection of these lines, this experimental design is not adequate to reveal the nature of the genes responsible for the movement into and out of the primed state. It was of interest, nonetheless, that high apoptosis lines consistently resulted from transfections with the human c-*myc* proto-oncogene in constitutive expression vectors.

4. GENETIC REGULATION

In more definitive experiments, candidate genes for regulation of apoptosis have been inserted into cultured cells under control of inducible promoters (Yonish-Rouach *et al.* 1991). Transgenic animals have been constructed in which the genes under test are expressed in a tissue-specific manner (McDonnell *et al.* 1989; Strasser *et al.* 1991). There are now several types of mice constitutively disabled in respect of particular genes, by germ-line mutations introduced through exposure to mutagens (Watanabe-Fukunaga *et al.* 1992) or as a result of engineered homologous recombination events (i.e. 'gene knockout' mice) (Clarke *et al.* 1992, 1993; Lowe *et al.* 1993*a*; Veis *et al.* 1993). These powerful techniques, sometimes applied together, are providing an increasingly detailed picture of the role of familiar oncogenes and oncosuppressor genes in the regulation of cell death.

Many of these developments will be described in later articles, but three generalizations are made here. First, expression of the proto-oncogene c-*myc* renders cells susceptible to apoptosis. This is true of cells of fibroblast (Evan *et al.* 1992; Fanidi *et al.* 1992; Bissonnette *et al.* 1992) and myeloid (Askew *et al.* 1991) lineages, although some lymphoid cell lines show rather different features (Yuh & Thompson 1989). As c-*myc* expression also sustains movement around the cell proliferation cycle, one attractive explanation of this finding is that c-*myc* induces a state in which both cell proliferation and apoptosis become possible, the critical choice between them depending upon additional considerations, such as the availability of growth factors (Evan *et al.* 1992). This dependence upon c-*myc* therefore emphasizes the existence of a high turnover state, in which cell proliferation and cell death are likely to coexist, their relative quantities being determined by the microenvironment. The striking coincidence of proliferation and death within the same areas in many tissues gives credibility to this view.

The second generalization is that activity of certain

genes protects cells in this high turnover state from apoptosis. The products of many of these genes are *survival factors*; they are not necessarily mitogens, although the surviving cells may of course proliferate if suitable conditions exist. Examples of such survival factors are the *bcl-2* family (Vaux *et al.* 1988) and the tyrosine kinase *abl* (Evans *et al.* 1993), the 55 kDa protein encoded by the adenovirus early region gene E1b (White *et al.* 1992), and the EBV transforming protein LMP-1 (Gregory *et al.* 1991), which may act in part by induction of host cell *bcl-2* (Henderson *et al.* 1991). All these survival factors confer resistance on cells that were previously sensitive to apoptosis. The resistance is pleiotropic in that it applies to a variety of pharmacologically diverse agents. Activity of the oncosuppressor gene *rb-1* also protects tissue cells from apoptosis (Clarke *et al.* 1992), but this may be fundamentally different from the action of survival factors as it is almost certainly not compatible with coexistent cell proliferation. The *rb*-dependent, apoptosis-resistant state may correspond to the growth arrest state of fibroblasts with down-regulated c-*myc*.

The third generalization is that cell injury – and in particular injury that causes DNA double-strand breaks – initiates apoptosis through p53. This was first demonstrated by insertion of a temperature-sensitive mutant of p53 into an IL-3 dependent myeloid cell line (Yonish-Rouach *et al.* 1991). At 37 °C, when p53 was expressed in mutant (oncogenic) conformation, the cells proliferated, whereas at 32 °C, when p53 was expressed in wild-type conformation, the cells underwent apoptosis. Later, it was shown that this effect of wild-type p53 applies at physiological levels of expression, since cells from animals expressing the normal two copies of wild-type p53 differ in their radiation sensitivity from cells of animals lacking any functional p53 gene (Lowe *et al.* 1993*a*; Clarke *et al.* 1993). Moreover, cells from heterozygotes, in which there is only a single copy of p53 show intermediate effects. Lack of p53 leads to a remarkable, total loss of apoptosis in response to ionizing radiation, even at high doses (14 Gy). This effect is found in many cell lineages: thymic T cells (Lowe *et al.* 1993*a*; Clarke *et al.* 1993), myeloid precursor cells (Lotem & Sachs 1993), intestinal epithelial cells (Merritt *et al.* 1994; Clarke *et al.* 1994) and in results still to be published, lymph node T cells and marrow pre-B cells. Although fibroblasts do not show p53-dependent apoptosis after radiation, their capacity to initiate apoptosis is revealed when they are manipulated to express the adenovirus E1a transforming protein (Lowe *et al.* 1993*b*).

5. HYPOTHESES AND IMPLICATIONS FOR DISEASE

These generalizations lead to new hypotheses relating to the regulation of cell number within tissues. It is probable that most growth factor stimuli reach tissue cells either through contact with other cells (whether homo- or heterotypic) or from the extracellular matrix. By this means local microenvironments are set up in which such stimuli are available, and outside of which they are not. While within the supportive microenvironment, cells are free to survive and proliferate: c-*myc* is expressed and apoptosis is not induced. Cells that find themselves outside, however, may survive in growth arrest, if they down-regulate c-*myc*, or may be deleted by apoptosis if they do not. There is very little information on the factors that might influence the choice between these options. In populations such as those in the lymph node follicle centre during B-cell affinity maturation, it is clear that the 'outside' cells mostly die; in this case they are those that lose out in the competition for engagement with the antigen presented on the follicular dendritic cells (Liu *et al.* 1989). Some cells, however, may be earmarked for survival, even should they leave the supportive microenvironment, by virtue of induction of genes such as *bcl-2*, coding for survival factors. In the follicle centre model, these are the memory B cells. A corollary of this model is that the cells most sensitive to death as a result of DNA damage will be those most equipped for apoptosis and least protected by survival factors. As indicated above, this is indeed the situation in many tissues: the follicle centre, the haemopoietic cells of the bone marrow and the lower third of the intestinal crypt epithelium being outstanding examples.

There are also interesting implications for major disease processes, of which two examples, cancer and AIDS, will be considered here. First, what is the fate of cells that survive, following DNA injury, because they lack functional p53? Animals without functional p53 accumulate such cells in large numbers, even after doses of x-irradiation that cause many strand breaks and mutations. Such animals have a high risk of cancer development, the heterozygotes showing a wide spectrum of primary sites, the homozygotes mostly dying early from T cell thymomas (Purdie *et al.* 1994). In both cases, the tumours are usually aneuploid. It is a reasonable, although still strictly unproven assumption, that the extra cells – those that would have been deleted in a normal animal – are the source of these malignant tumours. In this way, cancer can be conceived of as a disease resulting from deficiency in apoptosis.

The second example is the depletion of T cells in HIV-1 infection. Here the critical lesion appears to be inappropriate deletion, instead of proliferation, of T cells during activation (Laurent-Crawford *et al.* 1991; Banda *et al.* 1992), and perhaps specifically of cells otherwise destined to engender memory cells during repeated T cell-mediated responses (Howie *et al.* 1994). In this way, capacity for specific responses to often-repeated infection is selectively eroded, providing an elegant explanation for the well-known susceptibility of AIDS patients to infection by organisms of low pathogenicity that are commonly present in the human environment.

REFERENCES

Ansari, B., Coates, P.J., Greenstein, B.D. & Hall, P.A. 1993 *In situ* end-labelling detects DNA strand breaks in apoptosis and other physiological and pathological states. *J. Path.* **170**, 1–8.

Arends, M.J. & Wyllie, A.H. 1991 Apoptosis: mechanisms and role in pathology. *Int. Rev. exp. Path.* **32**, 223–254.

Arends, M.J., McGregor, A.H., Toft, N.J., Brown, E.J.H. & Wyllie, A.H. 1993 Susceptibility to apoptosis is differentially regulated by *c-myc* and mutated *Ha-ras* oncogenes and is associated with endonuclease availability. *Br. J. Cancer* **68**, 1127–1133.

Askew, D., Ashmun, R., Simmons, B. & Cleveland, J. 1991 Constitutive *c-myc* expression in IL-3-dependent myeloid cell line suppresses cycle arrest and accelerates apoptosis. *Oncogene* **6**, 1915–1922.

Banda, N.K., Bernier, J., Kurahara, D.K., Kurrle, R., Haigwood, N., Sekaly, R.P. & Finkes, T.H. 1992 Crosslinking CD4 by human immunodeficiency virus gp120 primes T cells for activation-induced apoptosis. *J. exp. Med.* **176**, 1099–1106.

Barry, M.A. & Eastman, A. 1993 Identification of deoxyribonuclease II as an endonuclease involved in apoptosis. *Arch. Biochem. Biophys.* **300**, 440–450.

Bissonette, R.P., Echeverri, F., Mahboubi, A. & Green, D.R. 1992 Apoptotic cell death induced by *c-myc* is inhibited by *bcl-2*. *Nature, Lond.* **359**, 552–554.

Clarke, A.R., Maandag, E.R., van Roon, M., van der Lugt, N.M.T., van der Valk, M., Hooper, M.L., Berns, A. & te Riele, H. 1992 Requirement for a functional *rb-1* gene in murine development. *Nature, Lond.* **359**, 328–330.

Clarke, A.R., Purdie, C.A., Harrison, D.J., Morris, R.G. & Bird, C.C. 1993 Thymocyte apoptosis induced by *p53*-dependent and independent pathways. *Nature, Lond.* **362**, 849–852.

Clarke, A.R., Gledhill, S., Hooper, M.L., Bird, C.C. & Wyllie, A.H. 1994 *p53* dependence of early apoptotic and proliferative responses within the mouse intestinal epithelium following γ-irradiation. *Oncogene* **9**, 1767–1773.

Duvall, E., Wyllie, A.H. & Morris, R.G. 1985 Macrophage recognition of cells undergoing programmed cell death (apoptosis). *Immunology* **56**, 351–358.

Evan, G.I., Wyllie, A.H., Gilbert, C.S., Littlewood, T.D., Land, H., Brooks, M., Waters, C.M., Penn, L.Z. & Hancock, D.L. 1992 Induction of apoptosis in fibroblasts by *c-myc* protein. *Cell* **69**, 119–129.

Evans, C.A., Owen-Lynch, J., Whetton, A.D. & Dive, C. 1993 Activation of the abelson tyrosine kinase activity is associated with suppression of apoptosis in hemopoietic cells. *Cancer Res.* **53**, 1735–1738.

Fadok, V.A., Savill, J.S., Haslett, C., Bratton, D.L., Doherty, D.E., Campbell, P.A. & Henson, P.M. 1992 Different populations of macrophages use either the vitronectin receptor or the phosphatidylserine receptor to recognize and remove apoptosic cells. *J. Immunol.* **149**, 4029–4038.

Fanidi, A., Harrington, E.A. & Evan, G.I. 1992 Cooperative interaction between *c-myc* and *bcl-2* protooncogenes. *Nature, Lond.* **359**, 554–556.

Fesus, L., Davies, P.J.A. & Piacentini, M. 1991 Apoptosis: molecular mechanisms in programmed cell death. *Eur. J. Cell Biol.* **56**, 170–177.

Gaido, M.L. & Cidlowski, J.A. 1991 Identification, purification and characterisation of a calcium-dependent endonuclease (nuc 18) from apoptotic rat thymocytes (nuc 18 is not histone H_2B). *J. biol. Chem.* **266**, 18580–18585.

Gavrieli, Y., Sherman, Y., Ben-Sasson, S.A. 1992 Identification of programmed cell death *in situ* via specific labeling of nuclear DNA fragmentation. *J. Cell Biol.* **119**, 493–501.

Gorczyca, W., Bruno, S. & Darzynkiewicz, Z. 1992 DNA strand breaks occuring during apoptosis: their early *in situ* detection by the terminal deoxynucleotidyl transferase and nick translation assays and prevention by serine protease inhibitors. *Int. J. Oncol.* **1**, 639–648.

Gregory, C.D., Dive, C., Henderson, S., Smith, C.A., Williams, G.T., Gordon, J. & Rickinson, A.B. 1991 Activation of Epstein Barr virus latent genes protects human B cells from death by apoptosis. *Nature, Lond.* **349**, 612–614.

Henderson, S., Rowe, M., Gregory, C.D., Croom-Carter, D., Wang, F., Longnecker, R., Kieff, G. & Rickinson, A.B. 1991 Induction of *bcl-2* expression by Epstein Barr virus latent membrane protein 1 protects infected B cells from programmed cell death. *Cell* **65**, 1107–1115.

Houge, G., Doskeland, S.O., Boe, R. & Lanotte, M. 1993 Selective cleavage of 28S rRNA variable regions V3 and V13 in myeloid leukaemia cell apoptosis. *FEBS Lett.* **315**, 16–20.

Howie, S.E.M., Sommerfield, A.J., Gray, E. & Harrison, D.J. 1994 Peripheral T lymphocyte depletion by apoptosis after CD4 ligation *in vivo*: selective loss of $CD44^-$ and 'activating' memory T cells. *Clin. exp. Immunol.* **95**, 195–200.

Kerr, J.F.R., Wyllie, A.H. & Currie, A.R. 1972 Apoptosis: a basic biological phenomenon with wide-ranging implications in tissue kinetics. *Br. J. Cancer* **26**, 239–257.

Kerr, J.F.R. 1971 Shrinkage necrosis: a distinct mode of cellular death. *J. Path.* **105**, 13–20.

Laurent-Crawford, A.G., Krust, B., Muller, S., Riviere, Y., Rey-Cuille, M.A., Bechet, J.M., Montagnier, L. & Hovanessian, A.G. 1991 The cytopathic effect of HIV is associated with apoptosis. *Virology* **185**, 829–839.

Liu, Y.J., Joshua, D.E., Williams, G., Smith, C.A., Gordon, J. & MacLennan, I.C. 1989 Mechanism of antigen-driven selection in germinal centres. *Nature, Lond.* **342**, 929–931.

Lotem, J. & Sachs, L. 1993 Hematopoietic cells from mice deficient in wild-type *p53* are more resistant to induction of apoptosis by some agents. *Blood* **82**, 1092–1096.

Lowe, S.W., Schmitt, E.M., Smith, S.W., Osborne, B.A. & Jacks, T. 1993*a* *p53* is required for radiation-induced apoptosis in mouse thymocytes. *Nature, Lond.* **362**, 847–849.

Lowe, S.W., Ruley, H.E., Jacks, T. & Housman, D.E. 1993*b* *p53* dependent apoptosis modulates the cytotoxicity of anticancer agents. *Cell* **74**, 957–967.

McDonnell, T.J., Deane, N., Platt, M., Nunez, G., Jaeger, U., McKearn, J.P. & Korsmeyer, S.J. 1989 *bcl-2*-immunoglobulin transgenic mice demonstrate extended B-cell survival and follicular lymphoproliferation. *Cell* **57**, 79–88.

Merritt, A.J., Potten, C.S., Kemp, C.J., Hickman, J.A., Balmain, A., Lane, D.P., Hall, P.A. 1994 The role of *p53* in spontaneous and radiation-induced apoptosis in the gastrointestinal tract of normal and *p53*-deficient mice. *Cancer Res.* **54**, 614–617.

Montpetit, M.L., Lowlep, K.R., Tenniswood, M. 1986 Androgen-repressed messages in the rat ventral prostate. *Prostate* **8**, 25–36.

Peitsch, M.C., Polzar, B., Stephen, H., Crompton, T., MacDonald, H.R., Mannherz, H.G. & Tschopp, J. 1993 Characterisation of the endogenous deoxyribonuclease involved in nuclear DNA degradation during apoptosis (programmed cell death) *EMBO J.* **12**, 371–377.

Purdie, C.A., Harrison, D.J., Peter, A., Dobbie, L., White, S., Howie, S.E.M., Salter, D.M., Bird, C.C., Wyllie, A.H., Hooper, M.L., Clarke, A.R. 1994 Tumour incidence, spectrum and ploidy in mice with a large deletion in the *p53* gene. *Oncogene* **9**, 603–609.

Roy, C., Brown, D.L., Little, J.E., Valentine, B.K., Walker, P.R., Sikorska, M., LeBlanc, J. & Chaly, N. 1992 The topoisomerase II inhibitor teniposide (VM-26) induces apoptosis in unstimulated mature murine lymphocytes. *Exp. Cell Res.* **200**, 416–424.

Saunders, J.W. 1966 Death in embryonic systems. *Science, Wash.* **154**, 604–612.

Savill, J.S., Dransfield, I., Hogg, N. & Haslett, C. 1990 Macrophage recognition of 'senescent self': the vitronectin receptor mediates phagocytosis of cells undergoing apoptosis. *Nature, Lond.* **342**, 170–173.

Shi, L., Kam, C.-M., Powers, J.C., Aebersold, R. & Greenberg, A.H. 1992 Purification of three cytotoxic lymphocyte granule serine proteases that induce apoptosis through distinct substrate and target cell interactions. *J. exp. Med.* **176**, 1521–1529.

Strasser, A., Harris, A.W., Cory, S. 1991 *Bcl-2* transgene inhibits T-cell death and perturbs thymic self-censorship. *Cell* **67**, 889–899.

Trump, B.F. & Berezesky, I.K. 1994 Cellular and molecular pathobiology of reversible and irreversible injury. In *Methods in toxicology*, vol. 1, part B (*In vitro toxicity indicators*) (ed. C. A. Tyson & J. M. Frazier), pp. 1–22. San Diego: Academic Press.

Vaux, D.L., Cory, S. & Adams, J.M. 1988 *Bcl-2* gene promotes haemopoietic cell survival and cooperates with *c-myc* to immortalise pre-B cells. *Nature, Lond.* **335**, 440–442.

Veis, D.J., Sorenson, C.M., Shutter, J.R. & Korsmeyer, S.J. 1993 *Bcl-2*-deficient mice demonstrate fulminant lymphoid apoptosis, polycystic kidneys, and hypopigmented hair. *Cell* **75**, 1–20.

Watanabe-Fukunaga, R., Brannan, C.I., Copeland, N.G., Jenkins, N.A. & Nagata, S. 1992 Lymphoproliferation disorder in mice explained by defects in *fas* antigen that mediates apoptosis. *Nature, Lond.* **356**, 314–317.

White, E., Sabbatini, P., Debbas, M., Wold, S.W.M., Kusher, D.I. & Gooding, L.R. 1992 The 19-kilodalton adenovirus E1B transforming protein inhibits programmed cell death and prevents cytolysis by tumour necrosis factor α. *Molec. Cell Biol.* **12**, 2571–2580.

Wyllie, A.H., Kerr, J.F.R. & Currie, A.R. 1980 Cell death: the significance of apoptosis. *Int. Rev. Cytol.* **68**, 251–306.

Wyllie, A.H. 1980 Glucocorticoid-induced thymocyte apoptosis is associated with endogenous endonuclease activation. *Nature, Lond.* **284**, 555–556.

Yonish-Rouach, E., Resnitzky, D., Lotem, J., Sachs, L., Kimchi, A. & Oren, M. 1991 Wild type *p53* induces apoptosis of myeloid leukaemic cells that is inhibited by interleukin-6. *Nature, Lond.* **352**, 345–347.

Yuan, J., Shaham, S., Ledoux, S., Ellis, H.M. & Horvitz, H.R. 1993 The *C. elegans* cell death gene *ced-3* encodes a protein similar to mammalian interleukin-1β-converting enzyme. *Cell* **75**, 641–652.

Yuh, Y.S. & Thompson, E.B. 1989 Glucocorticoid effect on oncogene/growth gene expression in human T lymphoblastic cell line CCRF-CEM. Specific *c-myc* RNA suppression by dexamethasone. *J. biol. Chem.* **264**, 10904–10910.

2

The ins and outs of programmed cell death during *C. elegans* development

MICHAEL O. HENGARTNER AND H. ROBERT HORVITZ

Howard Hughes Medical Institute, Department of Biology, Room 56-629, Massachusetts Institute of Technology, 77 Massachusetts Avenue, Cambridge, Massachusetts 02139, U.S.A.

SUMMARY

During the development of the *C. elegans* hermaphrodite, 131 of the 1090 cells generated undergo programmed cell death. Genetic studies have identified mutations in 14 genes that specifically affect this process. These genes define a genetic pathway for programmed cell death in *C. elegans*. Two genes, *ced-3* and *ced-4*, are required for cells to undergo programmed cell death, while a third gene, *ced-9*, protects cells that should live from undergoing programmed cell death. The proteins encoded by *ced-3* and *ced-9* show significant similarity to proteins that affect programmed cell death in vertebrates, suggesting that the molecular cell death pathway in which *ced-3*, *ced-4*, and *ced-9* act has been conserved between nematodes and vertebrates.

1. THE SETTING

The small nematode *Caenorhabditis elegans* has been widely used over the past two decades to study fundamental problems in developmental biology (Wood 1988). The two characteristics that have contributed most to the widespread use of this organism are its powerful genetics (Brenner 1974) and the simplicity and reproducibility of its development. For example, the pattern of cell divisions generating an adult worm from the one-celled zygote is essentially invariant from one animal to the next (Sulston & Horvitz 1977; Kimble & Hirsh 1979; Sulston *et al.* 1983).

During *C. elegans* development, a number of cells are eliminated, generally shortly after their generation, by undergoing programmed cell death. The number of cell deaths is high: in hermaphrodites, 131 of the 1090 cells generated die, while in males 147 of the 1178 cells generated die. Like the rest of *C. elegans* development, these deaths are highly reproducible: the same number of cells die in all individuals, and the time during development at which a given cell dies is constant. Programmed cell death in *C. elegans* can thus be studied with single-cell resolution.

2. THE CAST

Genetic studies of programmed cell death in *C. elegans* have led to the isolation of many mutations that affect this process. These mutations identify 14 genes that function in programmed cell death and define a genetic pathway for programmed cell death in *C. elegans* (figure 1). This pathway can be divided into four distinct steps: the decision of individual cells whether to undergo programmed cell death or adopt another fate, the actual killing of the cell, the engulfment of the dying cell by a neighboring cell, and the degradation of the dead, engulfed cell. Genes in the last three steps are involved in all programmed cell deaths, whereas genes in the decision step affect only a small number of cells, usually only one or two cell types.

3. THE PROTAGONISTS

(a) *The killers*

The activities of two genes, *ced-3* and *ced-4* (cell death abnormal), are necessary for programmed cell deaths to occur: mutations that inactivate either *ced-3* or *ced-4* result in the survival of almost all cells that normally die during development (Ellis & Horvitz 1986). The *ced-4* gene encodes a protein of about 63 kDa and is expressed primarily during embryonic development, when 113 of the 131 programmed cell deaths occur (Yuan & Horvitz 1992).

The ced-3 gene encodes a 503 amino acid protein and also is expressed primarily during embryonic development (Yuan *et al.* 1993). The protein encoded by the *ced-3* gene shows significant similarity to two mammalian proteins: interleukin-1β converting enzyme (ICE) and the product of the *nedd-2* gene (Yuan *et al.* 1993). ICE is a cysteine protease unrelated in sequence to any previously known protease (Cerretti *et al.* 1992; Thornberry *et al.* 1992). ICE recognizes and cleaves the 31 kDa pro-IL-1β into the 17.5 kDa mature IL-1β. The only other known ICE substrate besides pro-IL-1β is ICE itself: the protein is synthesized as an inactive proenzyme,

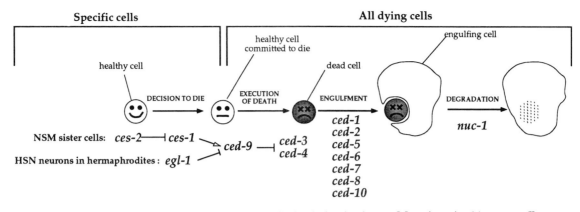

Figure 1. The genetic pathway for programmed cell death in *C. elegans*. Mutations in 14 genes affect programmed cell deaths. These mutations divide the process of programmed cell death into four steps; genes that act in the last three steps are common to all programmed cell deaths, whereas genes that act in the first step affect only a few cells. Regulatory interactions deduced from genetic studies are shown. ⟶, Positive regulation; ⟞, negative regulation. Adapted from Ellis *et al.* (1991).

which can be activated through limited proteolysis by the mature enzyme (Thornberry *et al.* 1992). The mouse *nedd-2* gene, which encodes a protein of unknown function, is expressed during embryonic brain development and is down-regulated in adult brain (Kumar *et al.* 1992). The sequence similarity between CED-3 and ICE suggests that CED-3 might also be a cysteine protease. Interestingly, overexpression of ICE, a vertebrate homologue of CED-3, induced programmed cell death in rat fibroblasts (Miura *et al.* 1993), suggesting that a cysteine protease similar to CED-3/ICE might also be involved in causing programmed cell death in mammals.

(b) *The guardian*

The activity of the gene *ced-9* appears to be both sufficient and necessary to protect *C. elegans* cells from undergoing programmed cell death. Either a gain-of-function mutation in the *ced-9* gene (Hengartner *et al.* 1992) or overexpression of wild-type *ced-9* (Hengartner & Horvitz 1994) results in the survival of cells that normally die. By contrast, mutations that inactivate *ced-9* cause many cells that normally survive to undergo programmed cell death (Hengartner *et al.* 1992). The progeny of animals lacking *ced-9* activity

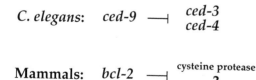

Figure 2. Nematodes and mammals might share a common pathway for programmed cell death. Genetic studies of *C. elegans* have indentified three genes, *ced-3*, *ced-4*, and *ced-9*, involved in the execution of all programmed cell death (top). Homologues of these genes might be involved in a similar pathway in mammals (bottom). Overexpression of CED-9 or Bcl-2 prevents programmed cell death (Vaux *et al.* 1988; Hengartner & Horvitz 1994), while overexpression of CED-3 or ICE induces programmed cell death (Miura *et al.* 1993).

die during embryogenesis, indicating that the protective activity of *ced-9* is essential for *C. elegans* development. The CED-9 protein shows sequence similarity to the product of the mammalian proto-oncogene *bcl-2* (Hengartner & Horvitz 1994). Interestingly, *bcl-2* also shares several functional similarities with *ced-9*: overexpression of *bcl-2* also protects cells from programmed cell death (apoptosis; reviewed by Korsmeyer *et al.* (1993)), and cells that lack *bcl-2* activity are hypersensitive to death-inducing signals (Nakayama *et al.* 1993; Veis *et al.* 1993). Human *bcl-2* can prevent both the normal programmed cell deaths that occur during *C. elegans* development (Vaux *et al.* 1992; Hengartner & Horvitz 1994) and also the ectopic cell deaths observed in mutants lacking *ced-9* function (Hengartner & Horvitz 1994), suggesting that *bcl-2* can substitute for *ced-9* in *C. elegans*. These results have led to the suggestion that *bcl-2* might be a vertebrate homologue of *ced-9*.

4. ACT I

How is the cell death program regulated such that only the correct cells die? The 131 cells that die during *C. elegans* hermaphrodite development have very different developmental origins and differ widely in their cell types, yet they all activate the same genetic pathway that leads to their programmed deaths. It is possible that like mammals (reviewed by Ellis *et al.* 1991), *C. elegans* has more than one way to induce programmed cell death and that distinct, possibly cell-type-specific death-inducing signals converge to activate the same pathway.

Several genes have been identified that affect the deaths of only a small number of cells. For example, the genes *ces-1* and *ces-2* (cell death specification) affect the decision of two cells in the pharynx (the worm's feeding organ) whether to live or die: in the wild-type, these two cells, which are the sister cells of the NSM motor neurons, undergo programmed cell death (Sulston *et al.* 1983). Dominant gain-of-function mutations in the gene *ces-1* and recessive loss-of-function mutations in the gene *ces-2* allow these two cells to survive and adopt a fate similar to that of their sisters the NSM neurons (Ellis & Horvitz 1991). A third gene, *egl-1* (egg-laying defective) affects the death of the two sexually dimorphic hermaphrodite-specific HSN neurons. The HSNs are a pair of serotonergic motor neurons that innervate the vulval muscles and drive egg-laying in *C. elegans* hermaphrodites (Desai *et al.* 1988; Desai & Horvitz 1989). In *C. elegans* males, which have no eggs to lay, the HSNs undergo programmed cell death (Sulston *et al.* 1983). Dominant mutations in the *egl-1* gene cause the HSNs to undergo programmed cell death in hermaphrodites, leading to a defect in egg-laying (Trent *et al.* 1983).

What is the function of these cell-death specification genes? One possibility is that they encode genes required for the cells to assume their proper identities: for example, the NSM sisters might survive in the *ces* mutants because they failed to adopt their proper fate and for this reason do not activate the cell death pathway. Alternatively, these genes might encode cell type-specific regulators of general cell death genes such as *ced-3*, *ced-4*, or *ced-9*. In this second model, the NSM sisters for example might be determined to die but fail to properly activate the cell death pathway, possibly as a result of the absence in the cell of one of the components of the general cell death pathway.

5. THE FINAL ACT

What happens once a cell has embarked upon the death process? Although the molecular events downstream of CED-3 and CED-4 are unknown, the structural changes accompanying death have been described (Sulston & Horvitz 1977; Robertson & Thomson 1982; Ellis *et al.* 1991). These changes include nuclear chromatin aggregation (and adoption by the nucleus of a characteristic pycnotic appearance evident at the levels of both the light and electron microscopes), cytoplasmic condensation, membrane whorling, and fragmentation of the cell into membrane-bound fragments. Several of these features are also characteristic of apoptotic deaths in mammals (Wyllie *et al.* 1980), suggesting that some of the molecular events responsible for these changes are common to both processes.

While these events occur within the dying cell, cytoplasmic extensions from a neighboring cell progressively surround the dying cell and engulf it. Mutations in seven genes (*ced-1, 2, 5, 6, 7, 8, 10*) delay or block this enfulfment process to varying degrees: whereas in the wild-type a dying cell is engulfed and degraded within an hour of the appearance of its first morphological changes, dying cells fail to be engulfed for many hours or even days in these mutants, leading to an accumulation of unengulfed, undegraded corpses that are easily observed in the light microscope (Hedgecock *et al.* 1983; Ellis *et al.* 1991).

The gene *nuc-1* (nuclease abnormal) is involved in the last step of the cell death pathway: *nuc-1* mutants lack a nuclease activity that is required to degrade the DNA of the dead cell. In these animals, cells die and are engulfed normally, but the DNA of the dead cells remains undegraded inside the engulfing cell. The enzymic properties of this nuclease (Hevelone & Hartman 1988) suggest that it corresponds to the major lysosomal nuclease, an observation consistent with the fact that this nuclease is required for the degradation of DNA of the bacteria on which the worm feeds (Hedgecock *et al.* 1983).

6. DEATH FROM THE INSIDE

Are programmed cell deaths suicides or murders? The discovery that the genes *ced-3* and *ced-4* are required for killing allowed this question to be addressed experimentally. Genetic experiments by Yuan & Horvitz (1990) suggest that both *ced-3* and *ced-4* genes must be expressed by the cells that die, indicating that the dying cell plays an essential role in bringing forth its own demise. This observation suggests that programmed cell death in *C. elegans* is a suicide process and truly comes 'from the inside'.

7. CHARACTER INTERPLAY

How do *ced-3*, *ced-4* and *ced-9* interact? Genetic studies showed that mutations in *ced-3* and *ced-4* can completely block the ectopic deaths caused by the lack of *ced-9* function (Hengartner *et al.* 1992). If we assume that these genes are involved in a regulatory pathway, then this observation suggests that *ced-9* encodes a negative regulator of *ced-3* and *ced-4* activities. The molecular mechanism of this regulation has yet to be uncovered. One possibility is that CED-9 is a protease inhibitor, binding to CED-3 to block its activity, or binding to the CED-3 pro-enzyme to prevent its proteolytic activation. Mammalian Bcl-2 binds to at least two distinct proteins, Bax (Oltvai *et al.* 1993) and R-ras (Fernandez-Sarabia & Bischoff 1993). Whether *C. elegans* contains homologues of these proteins to which CED-9 could bind remains to be determined.

8. AN OLD PLOT?

Two key players in the execution step, *ced-3* and *ced-9*, are members of gene families that include mammalian genes involved in the regulation of programmed cell death. For example, several members of the *ced-9* gene family, such as Bcl-2 (Vaux *et al.* 1988), Bax (Oltvai *et al.* 1993), and Bcl-x (Boise *et al.* 1993), modulate the susceptibility of mammalian cells to death-inducing signals, much in the same way as CED-9 does in *C. elegans*. Similarly, overexpression of ICE, a vertebrate homologue of CED-3, induced programmed cell death in rat fibroblasts (Miura *et al.* 1993), suggesting that a cysteine protease similar to CED-3 and ICE might cause programmed cell death in mammals.

The involvement of members of the *ced-9/bcl-2* and *ced-3*/ICE gene families in programmed cell death in both nematodes and mammals suggests that not only these genes but also the rest of the cell death pathway that has been characterized in *C. elegans* may be conserved through evolution and that such a pathway for programmed cell death is of ancient origin and might operate in all metazoans (figure 2).

We thank members of our laboratory for stimulating discussions. Work done in this laboratory was supported by the United States Public Health Service and by the Howard Hughes Medical Institute. M.O.H. was supported by a 1967 Science and Engineering Fellowship from the Natural Sciences and Engineering Research Council of Canada, a B1 Fellowship from the Fonds pour la Formation de Chercheurs et l'Aide à la Recherche of Québec, and a fellowship from the Glaxo Research Institute. H.R.H. is an Investigator of the Howard Hughes Medical Institute.

REFERENCES

Boise, L.H., Gonzáles-garcía, M., Postema, C.E. *et al.* 1993 *bcl-x*, a *bcl-2*-related gene that functions as a dominant regulator of apoptotic cell death. *Cell* **74**, 597–608.

Brenner, S. 1974 The genetics of *Caenorhabditis elegans*. *Genetics* **77**, 71–94.

Cerretti, D.P., Kozlosky, C.J., Mosley, B. *et al.* 1992 Molecular cloning of the interleukin-1β converting enzyme. *Science, Wash.* **256**, 97–100.

Desai, C., Garriga, G., McIntire, S.L. & Horvitz, H.R. 1988 A genetic pathway for the development of the *Caenorhabditis elegans* HSN motor neurons. *Nature, Lond.* **336**, 638–646.

Desai, C. & Horvitz, H.R. 1989 *Caenorhabditis elegans* mutants defective in the functioning of the motor neurons responsible for egg laying. *Genetics* **121**, 703–721.

Ellis, H.M. & Horvitz, H.R. 1986 Genetic control of programmed cell death in the nematode *C. elegans*. *Cell* **44**, 817–829.

Ellis, R.E. & Horvitz, H.R. 1991 Two *C. elegans* genes control the programmed deaths of specific cells in the pharynx. *Development* **112**, 591–603.

Ellis, R.E., Jacobson, D.M. & Horvitz, H.R. 1991 Genes required for the engulfment of cell corpses during programmed cell death in *Caenorhabditis elegans*. *Genetics* **129**, 79–94.

Ellis, R.E., Yuan, J. & Horvitz, H.R. 1991 Mechanisms and functions of cell death. *A. Rev. Cell Biol.* **7**, 663–698.

Fernandez-Sarabia, M.J. & Bischoff, J.R. 1993 Bcl-2 associates with the ras-related protein R-*ras* p23. *Nature, Lond.* **366**, 274–275.

Hedgecock, E.M., Sulston, J.E. & Thomson, J.N. 1983 Mutations affecting programmed cell deaths in the nematode *Caenorhabditis elegans*. *Science, Wash.* **220**, 1277–1279.

Hengartner, M.O., Ellis, R.E. & Horvitz, H.R. 1992 *C. elegans* gene *ced-9* protects cells from programmed cell death. *Nature, Lond.* **356**, 494–499.

Hengartner, M.O. & Horvitz, H.R. 1994 *C. elegans* cell death gene *ced-9* encodes a functional homolog of mammalian proto-oncogene *bcl-2*. *Cell* **76**, 665–676.

Hevelone, J. & Hartman, P.S. 1988 An endonuclease from *Caenorhabditis elegans*: partial purification and characterization. *Biochem. Genet.* **26**, 447–461.

Kimble, J. & Hirsh, D. 1979 The postembryonic cell lineages of the hermaphrodite and male gonads in *Caenorhabditis elegans*. *Devl Biol.* **70**, 396–417.

Korsmeyer, S.J., Shutter, J.R., Veis, D.J., Merry, D.E. & Oltvai, Z.N. 1993 Bcl-2/Bax: a rheostat that regulates an anti-oxidant pathway and cell death. *Semin. Cancer Biol* **4**, 327–332.

Kumar, S., Tomooka, Y. & Noda, M. 1992 Identification of a set of genes with developmentally down-regulated expression in the mouse brain. *Biochem. biophys. Res. Commun.* **185**, 1155–1161.

Miura, M., Zhu, H., Rotello, R., Hartwieg, E.A. & Yuan, J. 1993 Induction of apoptosis in fibroblasts by IL-1β converting enzyme, a mammalian homolog of the *C. elegans* cell death gene *ced-3*. *Cell* **75**, 653–660.

Nakayama, K., Nakayama, K., Negishi, I. *et al.* 1993 Disappearance of the lymphoid system in Bcl-2 homozygous mutant chimeric mice. *Science, Wash.* **261**, 1584–1588.

Oltvai, Z. N., Milliman, C. L. & Korsmeyer, S. J. 1993 Bcl-2 heterodimerizes *in vivo* with a conserved homolog, Bax, that accelerates programmed cell death. *Cell* **74**, 609–619.

Robertson, A. & Thomson, N. 1982 Morphology of programmed cell death in the ventral nerve cord of *Caenorhabditis elegans* larvae. *J. Embryol. exp. Morph.* **67**, 89–100.

Sulston, J.E. & Horvitz, H.R. 1977 Post-embryonic cell lineages of the nematode, *Caenorhabditis elegans*. *Devl Biol.* **56**, 110–156.

Sulston, J.E., Schierenberg, E., White, J.G. & Thomson, J.N. 1983 The embryonic cell lineage of the nematode *Caenorhabditis elegans*. *Devl Biol.* **100**, 64–119.

Thornberry, N.A., Bull, H.G., Calaycay, J.R. *et al.* 1992 A novel heterodimeric cysteine protease is required for interleukin-1β processing in monocytes. *Nature, Lond.* **356**, 768–774.

Trent, C., Tsung, N. & Horvitz, H.R. 1983 Egg-laying defective mutants of the nematode *Caenorhabditis elegans*. *Genetics* **104**, 619–647.

Vaux, D.L., Cory, S. & Adams, J.M. 1988 *bcl-2* gene promotes haemopoietic cell survival and cooperates with c-*myc* to immortalize pre-B cells. *Nature, Lond.* **335**, 440–442.

Vaux, D.L., Weissman, I.L. & Kim, S.K. 1992 Prevention of programmed cell death in *Caenorhabditis elegans* by human *bcl-2*. *Science, Wash.* **258**, 1955–1957.

Veis, D.J., Sorenson, C.M., Shutter, J.R. & Korsmeyer, S.J. 1993 Bcl-2-deficient mice demonstrate fulminant lymphoid apoptosis, polycystic kidneys, and hypopigmented hair. *Cell* **75**, 229–240.

Wyllie, A.H., Kerr, J.F.R. & Currie, A.R. 1980 Cell death: the significance of apoptosis. *Int. Rev. Cytol.* **68**, 251–306.

Yuan, J. & Horvitz, H.R. 1990 The *Caenorhabditis elegans* genes *ced-3* and *ced-4* act cell autonomously to cause programmed cell death. *Devl Biol.* **138**, 33–41.

Yuan, J. & Horvitz, H.R. 1992 The *Caenorhabditis elegans* cell death gene *ced-4* encodes a novel protein and is expressed during the period of extensive programmed cell death. *Development* **116**, 309–320.

Yuan, J., Shaham, S., Ledoux, S., Ellis, H.M. & Horvitz, H.R. 1993 The *C. elegans* cell death gene *ced-3* encodes a protein similar to mammalian interleukin-1beta converting enzyme. *Cell* **75**, 641–652.

3
Programmed cell death in *Drosophila*

H. STELLER, J. M. ABRAMS†, M. E. GRETHER AND K. WHITE‡

Howard Hughes Medical Institute, Department of Brain and Cognitive Sciences and Department of Biology, Massachusetts Institute of Technology, Cambridge, Massachusetts 02139, U.S.A.

SUMMARY

During *Drosophila* development, large numbers of cells undergo natural cell death. Even though the onset of these deaths is controlled by many different signals, most of the dying cells undergo common morphological and biochemical changes that are characteristic of apoptosis in vertebrates. We have surveyed a large fraction of the *Drosophila* genome for genes that are required for programmed cell death by examining the pattern of apoptosis in embryos homozygous for previously identified chromosomal deletions. A single region on the third chromosome (in position 75C1,2) was found to be essential for all cell deaths that normally occur during *Drosophila* embryogenesis. We have cloned the corresponding genomic DNA and isolated a gene, *reaper*, which is capable of restoring apoptosis when reintroduced into cell death defective deletions. The *reaper* gene is specifically expressed in cells that are doomed to die, and its expression precedes the first morphological signs of apoptosis by 1–2 h. This gene is also rapidly induced upon X-ray irradiation, and *reaper* deletions offer significant protection against radiation-induced apoptosis. Our results suggest that *reaper* represents a key regulatory switch for the activation of apoptosis in response to a variety of distinct signals.

In *Drosophila*, naturally occurring or programmed cell death (PCD) is a prominent phenomenon throughout most of the organism's life cycle. Large numbers of cells die both during embryonic development (Abrams *et al.* 1993; White *et al.* 1994) and later during metamorphosis (Kimura & Truman 1990), but some PCD continues into adult life (see, for example, Giorgi & Deri 1976). The decision about which cells will die is often not predetermined by lineage but, like in vertebrates, can be controlled by a variety of distinct epigenetic factors. This plasticity is particularly evident in the developing insect nervous system, where the onset of PCD can be regulated by cell–cell interactions (Cagan & Ready 1989; Wolff & Ready 1991; Ramos *et al.* 1993), trophic control (Steller *et al.* 1987; Campos *et al.* 1992) and the steroid hormone ecdysone (reviewed in Truman 1984). PCD in *Drosophila* appears to be important as a sculpting force during morphogenesis, to eliminate cells that have been damaged or are unable to complete their differentiation programme (for examples, see Fristrom 1969; Bryant 1988; Magrassi & Lawrence 1988; Dura *et al.* 1987; Smouse & Perrimon 1990; Abrams *et al.* 1993), and for establishing and maintaining the appropriate ratios among different cell types in organs that naturally vary in size. For example, the size of the *Drosophila* optic ganglia is always perfectly matched to the variable size of the eye (Power 1943). This matching in the number of cells is accomplished by adjusting the rate of both cell death as well as cell proliferation and differentiation through competitive interactions between retinal neurons and their targets (Fischbach & Technau 1984; Selleck & Steller 1991; Campos *et al.* 1991; Winberg *et al.* 1992).

Naturally occurring cell deaths in *Drosophila* typically display the characteristic morphological and biochemical changes associated with apoptosis in vertebrates (Kerr *et al.* 1972), but subtle morphological differences have been noted as well in some instances (see, for example, Giorgi & Deri 1976; Bryant 1988; Abrams *et al.* 1993; Wolff & Ready 1991). These similarities indicate that at least some of the molecular components of the basic cell death programme have been evolutionary conserved, a notion that has received direct support from recent molecular studies in other species (e.g. Vaux & Weissmann 1992; Rabizadeh *et al.* 1993; Yuan *et al.* 1993; Hengartner & Horvitz; 1994, this volume). The powerful genetic and molecular biology techniques available in *Drosophila* make it an ideal system for elucidating the mechanism by which cells undergo apoptosis. The *Drosophila* embryo is particularly well suited for a genetic analysis of PCD. First, a substantial amount of apoptosis occurs during embryogenesis in a rather predictable pattern, and these deaths can be rapidly and reliably visualized in live preparations using the vital dye acridine orange (Abrams *et al.* 1993). Second, by screening embryos as opposed to advanced developmental stages for the lack of PCD, one does not have to make any assumptions about the

† Present address: Deptartment of Cell Biology & Neuroscience, University of Texas Southwestern Medical Center, 5323 Harry Hines Boulevard, Dallas, Texas 75235, U.S.A.

‡ Present address: Cutaneous Biology Research Center, Massachusetts General Hospital, Harvard Medical School, MGH East, Building 149, Charlestown, Massachusetts 02129, U.S.A.

viability of cell death defective mutants. Finally, a large fraction of the *Drosophila* genome can be quickly surveyed by examining the pattern of cell death in embryos homozygous for previously characterized chromosomal deletions. Although these deletions typically include genes essential for viability, the large maternal supply of household functions (Garcia-Bellido *et al.* 1983) permits development well beyond the stage at which cell death begins.

By making use of an extensive collection of chromosomal deletions, we were able to screen approximately 50% of the *Drosophila* genome for functions globally required for PCD. The majority of deletions did not significantly affect the amount of cell death in the embryo, and many other deletions were associated with excessive AO staining, presumably as the result of developmental defects. A few deletions produced markedly reduced levels of PCD. These may affect the process by which cells are selected to die in response to specific signals. In contrast to all other deletions, we found only a single region on the third chromosome (position 75C1,2) to be required for all cell deaths that normally occur in the *Drosophila* embryo (White *et al.* 1994). Embryos homozygous for Df(3L)H99, the smallest cell death defective deletion available in the 75C1,2 interval, contained many extra cells and did not undergo certain morphogenetic movements, but developed a segmented cuticle and began to move. However, these embryos failed to hatch into larvae, apparently due to a strong defect in head involution. Genetic mosaic analysis with H99 indicates that this deletion does not contain any genes that are generally required for cell proliferation, differentiation or survival. Homozygous H99 mutant eye clones contain the full repertoire of morphologically normal, terminally differentiated cell types. The only abnormality we have detected in these clones is the apparent presence of extra cells, consistent with a block of PCD that normally occurs during eye development (K. White & H. Steller, unpublished results). Interestingly, H99 mutants were not only deficient in normal cell death, but also offered significant protection against ectopic cell deaths, such as those induced by X-irradiation and in developmental mutants. Such a global blockade of PCD in response to many distinct death signals indicates that this deletion removes a component of central importance to the cell death programme.

We have cloned all the genomic DNA corresponding to the H99 interval and identified a gene, *reaper*, which appears to play a central control function for the initiation of PCD (White *et al.* 1994). Significant levels of apoptosis can be restored in H99 mutants upon re-introducing a genomic *reaper* clone by germ line transformation experiments. Furthermore, expression of a *reaper* cDNA also leads to the induction of apoptosis (see below). Sequence analysis indicates that *reaper* encodes a small peptide of only 65 amino acids which shares no significant homologies to other known proteins (White *et al.* 1994). The open reading frame (ORF) of *reaper* has been highly conserved in *D. simulans*, a close relative of *D. melanogaster*, indicating that this sequence functions indeed as a protein coding

region. The *reaper* ORF has also been significantly conserved in *D. virilis* (M. Grether, A. F. Lamblin, R. Jespersen & H. Steller, unpublished results), which is separated from *D. melanogaster* by 60 million years of evolution. Due to its novel sequence, the biochemical function of this peptide is currently unknown. However, we have good reasons to believe that *reaper* is not a cell death effector protein, i.e. that it is not part of the cell death machinery itself. This conclusion is derived from the observation that some cell death can be induced in *reaper* deletions upon radiation with high doses of X-rays. Significantly, the few cells that die under these circumstances have all the morphological and biochemical characteristics of apoptosis. This indicates that the basic cell death programme is intact, but cannot be readily activated in *reaper* deletions. Consistent with the idea that *reaper* plays a central control function for the activation of PCD, we find that *reaper* mRNA is specifically expressed in cells that are doomed to die, and that expression of *reaper* precedes the first morphological signs of apoptosis by 1–2 h. In addition, X-irradiation of embryos leads to rapid and massive ectopic expression of *reaper*, followed by widespread cell death (J. Abrams & H. Steller, unpublished observations). Finally, *reaper* expression appears to be activated in cells that fail to differentiate due to genetic defects, and *reaper* deletions protect against cell death under these conditions (J. Abrams & H. Steller, unpublished observations). Our results suggest that most, if not all PCDs in *Drosophila* occur by one common mechanism, and that multiple signaling pathways for the induction of cell death converge onto the *reaper* gene. According to our model, *reaper* expression would lead to the selective activation of cell death effector proteins that may be present but inactive in most, if not all cells (see Raff 1992; Raff *et al.* 1993, this volume).

Our model predicts that expression of *reaper* should be sufficient for the induction of apoptosis. We have tested this model by generating transgenic fly strains that express a *reaper* cDNA clone under the control of the heat inducible hsp70 heat shock promoter. Upon heat shock induction, high levels of ectopic cell death were induced in transgenic embryos, and cells which would normally live initiated PCD (K. White & H. Steller, unpublished observations). We conclude that *reaper* expression is sufficient for the induction of apoptosis, and we propose that *reaper* represents a key regulatory function to activate a ubiquitous cell death effector pathway (figure 1). In principle, *reaper* could exert such a function by directly activating cell death effector proteins. Alternatively, *reaper* may inactivate proteins that protect cells from PCD, such as bcl-2 and/or related genes (e.g. Vaux *et al.* 1988; Hockenberry *et al.* 1988; Hengartner & Horvitz 1994). In the nematode *C. elegans*, inactivation of ced-9, a member of the bcl-2 family, is sufficient for the activation of PCD (Hengartner *et al.* 1992). A resolution between these possibility will require the identification of target proteins for *reaper*.

Another unresolved question is how multiple signalling pathways converge onto the *reaper* gene. It

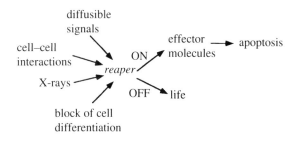

Figure 1. Model for the role of *reaper* during apoptosis. The induction of *reaper* appears to represent a key regulatory switch for the activation of a global cell death programme. Multiple distinct signalling pathways lead to the induction of *reaper* mRNA expression, and deletions that include *reaper* offer protection against apoptosis that is normally induced by these different signals. The expression of *reaper* is normally restricted to cells that will die, and ectopic expression of *reaper* is sufficient for the induction of apoptosis. Our model proposes that *reaper* acts upstream of cell death effector molecules, as the few cell deaths that occur in *reaper* deletions upon X-irradiation are morphologically indistinguishable from those in wild-type.

is conceivable that the integration of these pathways occurs at the level of the *reaper* promoter, by activation of transcription through the binding of distinct transcription factors. For example, in analogy to vertebrates (see Lane *et al.*, this volume), radiation may induce a p53 like protein in *Drosophila* which conceivably could bind to and activate *reaper* transcription. Alternatively, different signalling pathways could converge upstream of *reaper* and, in an extreme case, lead to the activation of *reaper* transcription through binding of a transcription factor to a single regulatory site. Finally, the specificity of *reaper* expression could be controlled at the post-transcriptional level, e.g. through selective RNA stabilization in cells that are doomed to die. Each of these models makes specific predictions that can be tested by standard promoter analyses.

A final point that warrants some discussion is the paucity of mutations that result in a global block of PCD in the *Drosophila* embryo. From screening a set of deletions covering approximately half of the *Drosophila* genome, we found only a single region to be required for all apoptotic deaths, and deleting this region did not even appear to affect the basic cell death machinery. On the other hand, there are several reasons to suspect that a number of genes are required for the onset and early stages of apoptosis, prior to the induction of acridine orange staining which was used as our initial assay. Acridine orange stains apoptotic cells only after nuclear and cytoplasmic condensation have become apparent, and actually appears to be a slightly later marker than TUNEL. In the nematode *C. elegans*, at least two genes, ced-3 and ced-4, are required for the onset of all PCDs in this organism (Ellis & Horvitz 1986, Ellis *et al.* 1991). There are several possible reasons why no other cell death genes have been identified in our screen. First, genes whose products are maternally supplied to the embryo would not have been picked up by us. However, it is not clear whether maternal products make a

significant contribution to apoptosis in *Drosophila*, because cell death begins only about 7 h after fertilization (Abrams *et al.* 1993), long after zygotic transcription has started. In our mind, a more likely explanation for the inability to identify deletions of cell death effector genes is the possibility that multiple genes may contribute to cell killing ('death from multiple causes'). If the absence of a cell death gene had simply resulted in slower deaths or deaths with a somewhat abnormal morphology, it would not have been picked up in our screen. However, it is precisely this kind of phenotype that may be expected from the removal of a cell death effector protein, and it should be possible to identify appropriate *Drosophila* mutants by more refined screens in the future. Alternatively, access to the elusive PCD effector proteins may be provided by isolating and characterizing proteins that interact with *reaper*. We expect that the combination of both genetic and biochemical approaches will make a major contribution to our understanding of the molecular basis of PCD. As many basic aspects of apoptosis and at least some of the proteins that regulate it have been evolutionary conserved, we expect that this work will also provide a better insight into the mechanisms of apoptosis in mammals, and its role in human pathogenesis.

We thank K. Farrell, R. Jespersen and L. Young for excellent technical assistance, and A. F. Lamblin for comments on this manuscript. This work was supported in part by a Pew Scholars Award (H.S.), and postdoctoral fellowships from the NIH (K.W.) and the ACS (J.M.A.). H.S. is an Associate Investigator with the Howard Hughes Medical Institute.

REFERENCES

Abrams, J., White, K., Fessler, L. and Steller, H. 1993 Programmed cell death during *Drosophila* embryogenesis. *Development* **117**, 29–43.

Bryant, P.J. 1988 Localized cell death caused by mutations in a *Drosophila* gene coding for a transforming growth factor-b homolog. *Devl Biol.* **128**, 386–395.

Cagan, R.L. and Ready, D.F. 1989 Notch is required for successive cell decisions in the developing *Drosophila* retina. *Genes Dev.* **3**, 1099–1112.

Campos, A.R., Fischbach, K.F. and Steller, H. 1992 Survival of photoreceptor neurons in the compound eye of *Drosophila* depends on connections with the optic ganglia. *Development* **114**, 355–366.

Campos-Ortega, J.A. and Hartenstein, V. 1985 *The embryonic development of Drosophila melanogaster.* Berlin: Springer-Verlag.

Dura, J., Randsholt, N.B., Deatrick, J., Erk, I., Santamaria, P., Freman, J.D., Freeman, S.J., Weddell, D. and Brock, H.W. 1987 A complex genetic locus, *polyhomeotic*, is required for segmental specification and epidermal development in *D. melanogaster*. *Cell* **51**, 829–839.

Ellis, H.M. and Horvitz, H.R. 1986 Genetic control of programmed cell death in *C. elegans*. *Cell* **44**, 817–829.

Ellis, R.E., Yuan, J.Y. and Horvitz, H.R. 1991 *A. Rev. Cell Biol.* **7**, 6–63.

James, A.A. and Bryant, P.J. 1981 Mutations causing pattern deficiencies and duplications in the imaginal wing disk of *Drosophila melanogaster*. *Devl Biol.* **85**, 39–54.

Fischbach, K.F. and Technau, G. 1984 Cell degeneration in the developing optic lobes of the sine oculis and small-optic-lobes mutants of *Drosophila melanogaster. Devl Biol.* **104**, 219–239.

Fristrom, D. 1969 Cellular degeneration in the production of some mutant phenotypes in *Drosophila melanogaster. Molec. gen. Genet.* **103**, 363–379.

Gavrielli, Y., Sherman, Y. and Ben-Sasson, S.A. 1992 Identification of programmed cell death *in situ* via specific labelling of nuclear DNA fragmentation. *J. Cell Biol.* **119**, 493–501.

Giorgi, F. and Deri, P. 1976 Cell death in the ovarian chanmbers of *Drosophila melanogaster. J. Embryol. exp. Morph.* **35**, 521–533.

Hengartner, M.O., Ellis, R.E. and Horvitz, H.R. 1992 *Caenorhabditis elegans* gene ced-9 protects cells from programmed cell death. *Nature, Lond.* **356**, 494–499.

Hengartner, M.O. and Horvitz, H.R. 1994 *C. elegans* cell survival gene ced-9 encodes a functional homolog of the mammalian proto-oncogene bcl-2. *Cell* **76**, 665–674.

Kimura, K. and Truman, J. W. 1990 Postmetamorphic cell death in the nervous and muscular systems of *Drosophila melanogaster. J. Neurosci.* **10**, 403–411.

Kerr, J.F.R., Wyllie, A.H. and Currie, A.R. 1972 Apoptosis: a basic biological phenomenon with wide ranging implications in tissue kinetics. *Br. J. Cancer* **26**, 239–257.

Magrassi, L. and Lawrence, P.A. 1988 The pattern of cell death in fushi tarazu, a segmentation gene of *Drosophila. Development* **104**, 447–451.

Power, M.E. 1943 The effect of reduction in numbers of ommatidia upon the brain of *Drosophila* melanogater. *J. exp. Zool.* **94**, 33–71.

Rabizadeh, S., La Count, D.J., Friesen, P.D. and Bredesen, D.E. 1993 Expression of the baculovirus p35 gene inhibits mammalian cell death. *J. Neurochem.* **61**, 2318–2321.

Raff, M.C. 1992 Social controls on survival and cell death. *Nature, Lond.* **356**, 397–400.

Raff, M.C., Barres, B.A., Burne, J.F., Coles, H.S., Ishizaki, Y. and Jacobson, M.D. 1993 Programmed cell death and the control of cell survival: lessons from the nervous system. *Science, Wash.* **262**, 695–670.

Ramos, R.G.P., Igloi, G.L., Lichte, B., Baumann, U., Maier, D., Schneider, T., Brandstätter, J.H., Fröhlich, A. and Fischbach, K.F. 1993 The *irregular chiasm C-roughest* locus of *Drosophila*, which affects axonal projections and programmed cell death, encodes a novel immunoglobulin-like protein. *Genes Dev.* **7**, 2533–2547.

Smouse, D. and Perrimon, N. 1990 Genetic dissection of a complex neurological mutant, polyhomeotic, in *Drosophila. Devl Biol.* **139**, 169–185.

Steller, H., Fischbach, K.F. and Rubin, G.M. 1987 Disconnected: a locus required for neuronal pathway formation in the visual system of *Drosophila. Cell* **50**, 1139–1153.

Truman, J.W. 1984 Cell death in invertebrate nervous systems. *A. Rev. Neurosci.* **7**, 171–188.

Wolff, T. and Ready, D.F. 1991 Cell death in normal and rough eye mutants of *Drosophila. Devl Biol.* **113**, 825–839.

White, K., Abrams, J., Grether, M., Young, L., Farrell, K. and Steller, H. 1994 Genetic control of cell death in *Drosophila. Science, Wash.* **264**, 677–683.

Winberg, M.L., Perez, S.E., and Steller, H. 1992 Generation and early differentiation of glial cells in the first optic ganglion of *Drosophila melanogaster. Development* **115**, 903–911.

Yuan, J. and Horvitz, H.R. 1990 The *Caenorhabditis elegans* genes ced-3 and ced-4 act cell autonomously to cause programmed cell death. *Devl Biol.* **138**, 33–41.

Yuan, J.Y., Shaham, S., Ledoux, S., Ellis, H.M. and Horvitz, H.R. 1993 The *C. elegans* cell death gene ced-3 encodes a protein similar to mammalian interleukin-1beta converting enzyme. *Cell* **75**, 641–652.

4

Block of neuronal apoptosis by a sustained increase of steady-state free Ca^{2+} concentration

JAMES L. FRANKLIN AND EUGENE M. JOHNSON JR

Department of Molecular Biology and Pharmacology, Washington University School of Medicine, 660 South Euclid Avenue, St Louis, Missouri 63110, U.S.A.

SUMMARY

Programmed death is a ubiquitous feature of the development of the vertebrate nervous system. This death is prevented *in vivo* by trophic factors and by afferent input. Death of neurons can also be prevented in culture models of programmed death by trophic factors and by chronic depolarization with elevated concentrations of K^+ in the culture medium. The latter effect is mediated by Ca^{2+} influx through voltage-gated channels and may prevent death by mimicking survival-promoting effects of naturally occurring electrical activity. Little is currently known about the mechanism by which either trophic factors or increased cytoplasmic Ca^{2+} promote survival.

1. INTRODUCTION

During the neurogenesis of the vertebrate nervous system, about fifty percent of neurons produced die at about the time functional connections to target tissues are being made (Oppenheim 1991). As this death is a normal component of development rather than a pathological process, it is called naturally occurring or programmed cell death (PCD). The apparent purpose of this death is to sculpt the developing nervous system by matching the number of neurons innervating a target tissue with the target size. Neurotrophic factors, released by target or other cells, are thought to be the principal agents controlling neuronal survival during and after PCD. Secreted in minute amounts, these factors bind to receptors on neuronal processes where they are internalized and retrogradely transported with the receptors to the neuronal soma. A retrogradely transported signal(s), which may be conveyed via the ligand–receptor complex, initiates mechanisms that maintain cellular survival and stimulates a variety of trophic responses. It is believed that neurons obtaining a sufficient amount of trophic factor survive, whereas those that are inadequately supplied undergo PCD.

Neuronal PCD resulting from insufficient availability of trophic factors can be mimicked *in vivo* by separating neurons from their targets via axotomy or target removal. As with naturally occuring PCD, the death caused by these insults can be averted by systemic administration of the appropriate trophic factor (Hendry & Campbell 1976; Hamburger *et al.* 1981) or, after axotomy, by administration of trophic factor at the lesion site (Rich *et al.* 1987). Systemic treatment of developing animals with neutralizing antibodies to growth factors also mimics PCD. Thus, neuronal PCD caused by inadequate amounts of trophic factors can be modelled *in vivo* via antibody administration, axotomy or target removal.

The physiological role of the prototypical neurotrophic factor, nerve growth factor (NGF), is the basis for the most studied neuronal death caused by insufficient availability of trophic factor. Depriving immature sympathetic (Levi-Montalcini & Booker 1960) or certain sensory neurons (Johnson *et al.* 1980) of NGF produces massive cell death *in vivo* and also in cell culture. NGF deprivation causes extensive sympathetic neuronal death even in adult animals indicating a continued requirement for its availability, although the rate of this death in older animals is considerably slower than in immature ones (Gorin & Johnson 1980). *In vivo*, NGF also blocks death caused by axotomy, target removal, and even certain chemical (Johnson & Aloe 1974) insults to these neurons. Thus, death of sympathetic and sensory neurons caused by NGF deprivation provides an appropriate physiological model for studying neuronal PCD. Dissociated embryonic rat superior cervical ganglion neurons (SCG) maintained in cell culture for 5–7 days in the presence of NGF hypertrophy and develop extensive neurites. Most of these neurons die within 48 h after being deprived of NGF. This death is preceded by condensation of chromatin, membrane blebbing, fragmentation of neurites, and atrophy of the cell soma. Prior to demise the DNA of NGF-deprived SCG neurons becomes fragmented into oligonucleosomes (Edwards *et al.* 1991; Deckwerth & Johnson 1993). Descriptions of morphological changes after trophic factor deprivation of other types of neurons are similar (see Pilar & Landmesser 1976). The morphological changes and DNA fragmentation that occur after neurotrophic factor deprivation suggests that the death of these cells is apoptosis, a type of death first identified in non-neuronal cells (Duvall & Wyllie 1986). Further evidence that the death of sympathetic neurons after trophic factor deprivation is apoptotic comes from experiments with inhibitors of macromolecular synthesis. In some types of cells, apoptosis appears to require production of new proteins since it is blocked by inhibitors of RNA and protein synthesis. Sympathetic neuronal PCD also exhibits this characteristic; block of macromolecular synthesis with inhibitors such as actinomycin-D and cycloheximide completely prevents death of these cells after NGF deprivation *in vitro* (Martin *et al.* 1988). Block of PCD by inhibition of macromolecular synthesis has also been demonstrated *in vivo* in motor neurons of chicken embryos (Oppenheim *et al.* 1990).

2. ELEVATED POTASSIUM BLOCKS NEURONAL PCD *IN VITRO*

Scott & Fisher (1970) found that elevated extracellular K^+ concentrations ($[K^+]_o$) can maintain the viability of embryonic chicken dorsal root ganglion neurons *in vitro* that would die in normal $[K^+]_o$ (about 5 mM in vertebrates). Subsequent work from several laboratories demonstrated a similar effect of high $[K^+]_o$ on the survival of many other types of neurons from both the peripheral and central nervous systems (Franklin & Johnson 1992). Figure 1 illustrates the ability of increased $[K^+]_o$ to promote the survival of rat SCG neurons in the absence of NGF. When these cells are maintained in culture by high $[K^+]_o$ they appear morphologically very similar to neurons maintained by NGF and will live and remain healthy for weeks or months without NGF. While other agents (such as those increasing cellular cAMP, for example) are capable of supporting survival of sympathetic neurons in the absence of NGF (Martin *et al.* 1992), none have as potent an effect on survival as high $[K^+]_o$. In fact, it appears that high $[K^+]_o$ can almost exactly substitute for NGF in supporting survival of sympathetic neurons in culture (Franklin *et al.* 1994).

Intracellular and extracellular K^+ concentrations are the primary determinants of the resting membrane potential of neurons. Increasing $[K^+]_o$ *in vitro* causes depolarization to a new resting membrane potential at which cells remain for the duration of the exposure to the elevated $[K^+]_o$ (Chalazonitis & Fischbach 1980). Such sustained depolarization may promote survival by imitating the effects of naturally occurring electrical activity, i.e. afferent input (Franklin & Johnson 1992). This idea is supported by numerous studies demonstrating that pharmacological blockade of electrical activity or removal of afferent input can increase the death of some populations of developing neurons (for review, see Oppenheim 1991). Nishi & Berg (1981) demonstrated that enhancement of the survival of chicken ciliary ganglion neurons by high $[K^+]_o$ is blocked by the relatively non-selective Ca^{2+} channel antagonists D-600 and Mg^{2+}, suggesting that influx of Ca^{2+} through voltage-gated channels is responsible for mediating the survival-promoting

effects of depolarization. More recent work has shown that promotion of the survival of several types of neurons by high $[K^+]_o$ can be prevented by the more specific dihydropyridine (DHP) Ca^{2+} channel antagonists (Gallo *et al.* 1987; Collins & Lile 1989; Koike *et al.* 1989). Conversely, at least some neurons are made more responsive to increased $[K^+]_o$ by DHP Ca^{2+}-channel agonists. These data suggest that, in these types of neurons, chronic depolarization causes activation of DHP-sensitive or L-type Ca^{2+} channels (Fox *et al.* 1987) and that influx of Ca^{2+} through these channels is responsible for the enhanced survival.

In addition to the pharmacological evidence, a role for voltage-gated Ca^{2+} channels in high $[K^+]_o$-promoted survival is supported by measurements of intracellular free Ca^{2+} concentrations ($[Ca^{2+}]_i$) with the Ca^{2+}-sensitive dye fura-2 (Grynkiewicz *et al.* 1985). Acute exposure of neurons to elevated $[K^+]_o$ causes a rapid elevation of $[Ca^{2+}]_i$. After the initial increase, $[Ca^{2+}]_i$ falls, in at least some types of neurons, to a new steady-state concentation that remains elevated above baseline levels for hours or days with continued exposure to increased $[K^+]_o$. A sustained rise of $[Ca^{2+}]_i$ in K^+-depolarized neurons has now been demonstrated in rat sympathetic neurons (Koike & Tanaka 1991; Franklin *et al.* 1994), chicken ciliary ganglion neurons (Collins *et al.* 1991), and rat cerebellar granule neurons (Bessho *et al.* 1994). Treatment of neurons with agents that block promotion of survival by chronic depolarization, such as DHP Ca^{2+}-channel antagonists, return the sustained rise of $[Ca^{2+}]_i$ to basal, or near basal, levels. Figure 2 illustrates the effects of elevated $[K^+]_o$ and the DHP antagonist, nifedipine, on steady-state $[Ca^{2+}]_i$ and survival of rat SCG neurons in culture. In these cells low concentrations of nifedipine block both the sustained rise of $[Ca^{2+}]_i$ caused by K^+ depolarization and also block survival. These data suggest that the sustained increase of $[Ca^{2+}]_i$ caused by influx of Ca^{2+} through DHP-sensitive, voltage-gated channels is responsible for promoting survival of these cells rather than some other effect of depolarization or high $[K^+]_o$.

In addition to chronic depolarization, we have found that NGF-deprived rat sympathetic neurons are saved from death by treatment with the sesquiterpene lactone, thapsigargin (Lampe *et al.* 1992), a potent and selective inhibitor of a Ca^{2+} pump responsible for sequestration of Ca^{2+} into a subset of intracellular Ca^{2+} stores (Thastrup *et al.* 1990). In many types of cells, exposure to thapsigargin causes a rapid rise of cytoplasmic $[Ca^{2+}]_i$ as Ca^{2+} leaks out of these stores after pump inhibiton. The plasma membrane Ca^{2+} pumps, which are unaffected by thapsigargin, then remove excess Ca^{2+} and reduce $[Ca^{2+}]_i$. However, $[Ca^{2+}]_i$ in some types of cells does not return to baseline concentrations after thapsigargin exposure but remains elevated at a new steady-state level. It is thought that a messenger molecule, released by depletion of the Ca^{2+} stores, activates a plasma membrane Ca^{2+} conductance and the resulting Ca^{2+} influx causes the sustained rise of $[Ca^{2+}]_i$

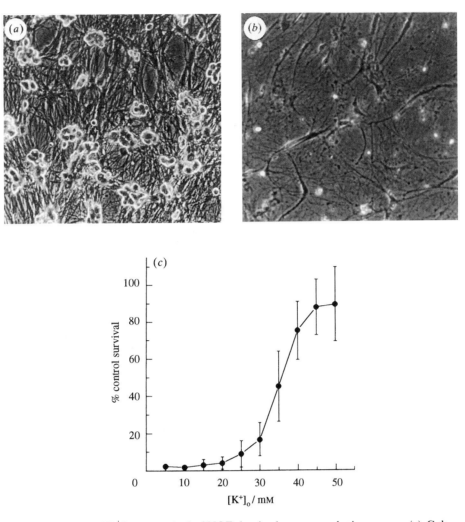

Figure 1. Effect of increasing $[K^+]_o$ on survival of NGF-deprived rat sympathetic neurons. (*a*) Culture of rat SCG neurons maintained for 6 days in medium containing an anti-NGF antibody and 50 mM extracellular K^+ with no NGF. Neurons are alive and healthy. (*b*) Culture maintained in medium with 5 mM extracellular K^+, but without NGF for 6 days. All neurons have died. (*c*) Quantification of the ability of different $[K^+]_o$ to maintain survival of these neurons in the absence of NGF. Cells were deprived of NGF 6 days after plating embryonic-day-21 cells and were exposed to the indicated $[K^+]_o$ for 4 days. Survival was assayed by blinded counting. Control survival is the number of cells maintained in NGF and 5 mM K^+ for the same period as the deprived cells. Error bars are s.d., $n = 8–12$ for each data point.

(Hoth & Penner 1992; Putney 1993). Our work shows that thapsigargin not only saves NGF-deprived sympathetic neurons from dying, but also causes a DHP-insensitive sustained increase of $[Ca^{2+}]_i$. Increasing extracellular Ca^{2+} in the presence, but not absence, of thapsigargin increases $[Ca^{2+}]_i$ to levels similar to those associated with optimal survival with $[K^+]_o$ but does not depolarize cells to membrane potentials that are associated with survival promotion by elevated $[K^+]_o$ (unpublished observation). Therefore, thapsigargin treatment provides a means for inducing a sustained rise of $[Ca^{2+}]_i$ that prevents PCD, is independent of depolarization, and independent of activation of L-type Ca^{2+} channels. This finding lends strong support to the notion that the elevation of steady-state $[Ca^{2+}]_i$ is responsible for promoting survival by high $[K^+]_o$.

Based upon our work and that of others, we have suggested a 'Ca^{2+} set-point' hypothesis of neuronal survival and neurotrophic factor dependence (Koike *et al*. 1989; Franklin & Johnson 1992; Johnson *et al*. 1992). This hypothesis is similar to that of Kater *et al*. (1988) regarding the effects of $[Ca^{2+}]_i$ on neurite outgrowth. The survival set-point hypothesis posits four steady-state levels, or set-points, of $[Ca^{2+}]_i$ that affect neuronal survival. The first set-point encompases levels of $[Ca^{2+}]_i$ that are too low to support essential Ca^{2+}-dependent processes and cause neuronal death even in the presence of appropriate neurotrophic factors. The existence of this set-point is supported by experiments showing that very low levels of $[Ca^{2+}]_i$ are deleterious to neuronal growth and survival in cell culture (Tolkovsky *et al*. 1990; J. L. Franklin & E. M. Johnson, unpublished observations). Although neurons can be killed *in vitro* by reducing $[Ca^{2+}]_i$ to low levels by incubating them in medium containing low Ca^{2+} concentrations, *in vivo*, these levels of $[Ca^{2+}]_i$ probably never occur and this set-point is, therefore, of little relevance to physiological situations. In the presence of

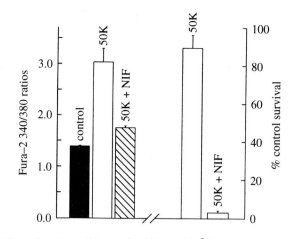

Figure 2. Effect of elevated $[K^+]_o$ on $[Ca^{2+}]_i$ and survival of scg neurons maintained in culture medium containing 50 mM K^+ (50 K) and no NGF. Chronic depolarization with high $[K^+]_o$ caused a sustained increase of steady-state $[Ca^{2+}]_i$ (shown as higher Fura-2 340/380 ratios). The DHP Ca^{2+} channel antagonist nifedipine largely blocked this increase. Cells were depolarized with 50 K for 24 h and exposed to 100 nM nifedipine for 3–4 h before measurements were made. $n = 25-27$ for each bar. Survival enhancement of NGF-deprived neurons caused by growth in 50 K medium was also blocked by 100 nM nifedipine. Cells were deprived of NGF on the fifth day after plating and maintained four additional days in 50 K ± nifedipine. Survival was then assayed by blinded counting. $n = 9-12$ for each bar.

neurotrophic factors, resting $[Ca^{2+}]_i$ (about 100 nM in many cells; the second set-point) is adequate for the survival of trophic-factor-dependent neurons. A modest sustained elevation of $[Ca^{2+}]_i$, such as that induced by elevated $[K^+]_o$ or thapsigargin treatment (100 nM to several 100 nM), converts neurons to trophic-factor-independence. Thus, $[Ca^{2+}]_i$, over a limited range, determines the trophic factor dependence of neurons. These levels of $[Ca^{2+}]_i$ comprise the third set-point of our hypothesis. This set-point is of possible physiological relevance as the electrical activity of neurons resulting from afferent input can increase $[Ca^{2+}]_i$ and may reduce the requirement for neurotrophic support from target tissues (Schmidt & Kater 1993). Very high levels of $[Ca^{2+}]_i$ such as those occurring during excitotoxicity stimulate injurious processes within cells and are toxic to neurons (Choi 1987, 1988). These toxic levels of $[Ca^{2+}]_i$ comprise the fourth set-point of the hypothesis. Therefore, the near absence of $[Ca^{2+}]_i$ (the first set-point) or an excess of $[Ca^{2+}]_i$ (the fourth set-point) cause neuronal death; the second set-point allows neuronal survival with support by neurotrophic factor, while the third set-point promotes survival of neurons independently from neurotrophic factor.

3. MECHANISMS BY WHICH INCREASED $[Ca^{2+}]_i$ PROMOTES SURVIVAL

Little is known about the mechanisms underlying survival-promotion by elevated $[Ca^{2+}]_i$. However, because intracellular free Ca^{2+} is a major second-messenger molecule, it is likely that increased $[Ca^{2+}]_i$ promotes survival by stimulating a signal transduction

pathway, possibly the same one stimulated by trophic factors. Identification of Ca^{2+} signalling pathways for survival may, therefore, also provide information about trophic-factor signalling. Calcium signals are translated into cellular events by intracellular Ca^{2+}-binding proteins. Of these proteins the most ubiquitous and abundant is calmodulin, the principal intracellular Ca^{2+} receptor of all non-muscle eukaryotic cells. When bound to Ca^{2+}, calmodulin modulates the activity of a number of Ca^{2+}-dependent molecules that are involved in a myriad of cellular events. Because of its abundance, ubiquity, and multiple effects, calmodulin is a reasonable candidate Ca^{2+}-binding protein for the initial signal transduction event that leads to Ca^{2+}-promoted survival. We have found that the calmodulin antagonists, calmidazolium and W7, prevent survival of depolarized NGF-deprived rat sympathetic neurons at concentrations that do not affect survival in the presence of NGF and normal $[K^+]_o$. However, these compounds appear to inhibit survival by blocking the sustained increase of Ca^{2+} caused by the chronic depolarization rather than through a specific effect on calmodulin (Franklin *et al.* 1992, 1994). The block of the sustained $[Ca^{2+}]_i$ increase is probably caused by antagonism of Ca^{2+} channels carrying the Ca^{2+} influx responsible for the increased $[Ca^{2+}]_i$ (Doroshenko *et al.* 1988). Recently, Hack *et al.* (1993) showed that calmodulin antagonists block depolarization-enhanced survival of embryonic rat cerebellar ganglion neurons at concentrations that do not significantly affect Ca^{2+} influx and that an antagonist of Ca^{2+}-calmodulin-dependent protein kinase (CAM-kinase) blocks depolarization-enhanced survival of these cells. These results suggest that a calmodulin-CAM-kinase pathway is responsible for the effect of Ca^{2+} on survival in these neurons. However, the effect on $[Ca^{2+}]_i$ of these compounds was not determined leaving open the possibility that the block of survival may be through an effect on $[Ca^{2+}]_i$ rather than on calmodulin. Indeed, preliminary experiments in our laboratory with fura-2 loaded granule cells have found that a survival-inhibiting concentration of calmidazolium (1 μM) blocks the sustained increase of $[Ca^{2+}]_i$ caused by 25 mM extracellular K^+ (T. Miller & E. M. Johnson, unpublished observations). We have not yet tested the effects of CAM-kinase inhibitors on $[Ca^{2+}]_i$ in these cells. Thus, while calmodulin may well be a mediator of Ca^{2+}-promoted neuronal survival, lack of specificity of available calmodulin inhibitors has hampered investigation of this possibility.

Another protein whose activity can be affected by increases of $[Ca^{2+}]_i$, protein kinase C (PKC), has been implicated in depolarization-promoted survival of embryonic chicken sympathetic neurons in culture. In these cells depolarization with elevated $[K^+]_o$ increases protein kinase C (PKC) activity and phorbol esters, which mimic PKC activation by diacyl glycerol, can substitute for chronic depolarization in promoting survival (Wakade *et al.* 1988). However, phorbol esters have no effect on survival of NGF-deprived rat sympathetic neurons (Martin *et al.* 1992); thus, PKC is probably not involved in depolarization-enhanced survival of these cells. It is also unlikely that Trk, the high-affinity NGF receptor, is a part of the signal-transduction pathway for depolarization/Ca^{2+}-promoted survival of sympathetic neurons. We have found that, in SCG neurons, NGF causes constitutive phosphorylation of Trk on tyrosine residues, an indication of the activation of Trk tyrosine kinase activity (Franklin *et al.* 1992). Removal of NGF causes this phosphorylation to drop to undetectable levels. After deprivation, NGF readdition causes rapid rephosphorylation of Trk but depolarization has no effect. Therefore, it appears that Ca^{2+} and NGF signalling, if they converge on the same survival-signalling pathway do so downstream of Trk. A possible point of convergence of the two survival promoting signals is MAP kinases. In PC12 cells, MAP kinases show increased tyrosine phosphorylation when the cells are exposed to either NGF or elevated $[K^+]_o$ (Tsao *et al.* 1990). In total cell lysates of rat SCG neurons, we have found that a $44\,kDa$ protein, probably a MAP kinase, is tyrosine phosphorylated by NGF and chronic depolarization (Franklin *et al.* 1994). It will be of interest to study further the effects of trophic factors and depolarization on the activity of these kinases and to characterize the downstream effects of their activation that may be involved in promotion of survival.

When embryonic rat SCG neurons are maintained in culture in the presence of NGF, they not only survive but exhibit sustained growth for at least several weeks after initial plating. The cell soma hypertrophies, neurite outgrowth occurs at a linear rate, and total protein content increases at a linear rate. When these cells are maintained in the absence of NGF with optimal survival-promoting $[K^+]_o$ little, if any, somatic hypertrophy, neurite outgrowth, or increase of protein content occurs. Therefore, although removal of NGF in the presence of optimal survival-promoting $[K^+]_o$ (50 mM) maintains these cells in a viable state, growth almost completely ceases. Thus, while NGF promotes both growth and survival of rat sympathetic neurons in culture, chronic depolarization supports only survival (Franklin *et al.* 1994). This finding strongly suggests that survival and growth are separable process mediated by different signalling pathways. There currently is considerable interest in utilizing neurotrophic factors for treatment of neurodegenerative diseases (Olson 1993). It is likely that a secondary consequences of such treatment is unwanted outgrowth of neurites. Because chronic depolarization stimulates survival-promotion, but not growth, it may be a useful means of teasing out the survival signal transduction pathway(s) from the signalling pathway(s) that induce growth. Such data would may aid in the development of agents that could be used clinically to enhance neuronal survival without causing ectopic growth.

REFERENCES

Bessho, Y., Nawa, H. & Nakanishi, S. 1994 Selective up-regulation of an NMDA receptor subunit mRNA in cultured cerebellar granule cells by K^+-induced depolarization and NMDA treatment. *Neuron* **12**, 87–95.

Chalazonitis, A. & Fischbach, G.D. 1980 Elevated potassium induces morphological differentiation of dorsal root ganglionic neurons in dissociated cell culture. *Devl Biol.* **78**, 172–183.

Choi, D.W. 1987 Ionic dependence of glutamate neurotoxicity. *J. Neurosci.* **7**, 369–379.

Choi, D.W. 1988 Calcium-mediated neurotoxicity: relationship to specific channel types and role in ischemic damage. *Trends Neurosci.* **11**, 465–469.

Collins, F. & Lile, J.D. 1989 The role of dihydropyridine-sensitive voltage-gated calcium channels in potassium-mediated neuronal survival. *Brain Res.* **502**, 99–108.

Collins, F., Schmidt, M.F., Guthrie, P.B. & Kater, S.B. 1991 Sustained increase in intracellular calcium promotes neuronal survival. *J. Neurosci.* **11**, 2582–2587.

Deckwerth, T.L. & Johnson, E.M. Jr 1993 Temporal analysis of events associated with programmed cell death (apoptosis) of sympathetic neurons deprived of nerve growth factor. *J. Cell Biol.* **123**, 1207–1222.

Doroshenko, P.A., Kostyuk, P.G. & Luk'yanetz, E.A. 1988 Modulation of calcium currents by calmodulin antagonists. *Neuroscience* **27**, 1073–1080.

Duvall, E. & Wyllie, A.H. 1986 Death and the cell. *Immunol. Today* **7**, 115–119.

Edwards, S.N., Buckmaster, A.E. & Tolkovsky, A.M. 1991 The death programme in cultured sympathetic neurones can be suppressed at the posttranslational level by nerve growth factor, cyclic AMP, and depolarization. *J. Neurochem.* **57**, 2140–2143.

Fox, A.P., Nowycky, M.C. & Tsien, R.W. 1987 Kinetic and pharmacological properties distinguishing three types of calcium currents in chick sensory neurones. *J. Physiol.*, Lond. **394**, 149–172.

Franklin, J.L. & Johnson, E.M. Jr 1992 Suppression of programmed neuronal death by sustained elevation of cytoplasmic calcium. *Trends Neurosci.* **15**, 501–508.

Franklin, J.L., Juhasz, A., Cornbrooks, E.B., Lampe, P.A. & Johnson, E.M. Jr 1992 Promotion of sympathetic neuronal survival by chronic depolarization and increased $[Ca^{2+}]_i$ is a threshold phenomenon. *Soc. Neurosci. Abstr.* **18**, 30.4.

Franklin, J.L., Sanz-Rodriguez, C., Juhasz, A., Deckwerth, T.L. & Johnson, E.M. Jr 1994 Chronic depolarization prevents programmed death of sympathetic neurons *in vitro* but does not support growth: requirement for Ca^{2+} influx but not Trk activation. (Submitted.)

Gallo, V., Kingsbury, A., Balázs, R. & Jørgensen, O.S. 1987 The role of depolarization in the survival and differentiation of cerebellar granule cells in culture. *J. Neurosci.* **7**, 2203–2213.

Gorin, P.D. & Johnson, E.M. Jr 1980 Effects of long-term nerve growth factor deprivation on the nervous system of the adult rat: an experimental autoimmune approach. *Brain Res.* **198**, 27–42.

Grynkiewicz, G., Poenie, M. & Tsien, R.T. 1985 A new generation of Ca^{2+} indicators with greatly improved fluorescence properties. *J. biol. Chem.* **260**, 3440–3450.

Hack, N., Hidaka, H., Wakefield, M.J. & Balázs, R. 1993 Promotion of granule cell survival by high K^+ or excitatory amino acid treatment and Ca^{2+}/calmodulin-dependent protein kinase activity. *Neuroscience* **57**, 9–20.

Hamburger, V., Brunso-Bechthold, J.K. & Yip, J.W. 1981 Neuronal death in the spinal ganglia of the chick embryo and its reduction by nerve growth factor. *J. Neurosci.* **1**, 60–71.

Hendry, I.A. & Campbell, J. 1976 Morphometric analysis of rat superior cervical ganglion after axotomy and nerve growth factor treatment. *J. Neurocytol.* **5**, 351–360.

Hoth, M. & Penner, R. 1992 Depletion of intracellular

calcium stores activates a calcium current in mast cells. *Nature, Lond.* **355**, 353–355.

Johnson, E.M. Jr & Aloe, L. 1974 Suppression of the in vitro and in vivo cytotoxic effects of guanethidine in sympathetic neurons by nerve growth factor. *Brain Res.* **81**, 519–532.

Johnson, E.M. Jr, Gorin, P.D., Brandeis, L.D. & Pearson, J. 1980 Dorsal root ganglion neurons are destroyed by exposure in utero to maternal antibody to nerve growth factor. *Science, Wash.* **210**, 916–918.

Johnson, E.M. Jr, Koike, T. & Franklin, J. 1992 A "calcium set-point hypothesis" of neuronal dependence on neurotrophic factor. *Expl Neurol.* **115**, 163–166.

Kater, S.B., Mattson, M.P., Cohan, C. & Connor, J. 1988 Calcium regulation of the neuronal growth cone. *Trends Neurosci.* **11**, 315–321.

Koike, T. & Tanaka, S. 1991 Evidence that nerve growth factor dependence of sympathetic neurons for survival in vitro may be determined by levels of cytoplasmic free Ca^{2+}. *Proc. natn. Acad. Sci. U.S.A.* **88**, 3892–3896.

Koike, T., Martin, D.P. & Johnson, E.M. Jr 1989 Role of Ca^{2+} channels in the ability of membrane depolarization to prevent neuronal death induced by trophic-factor deprivation: Evidence that levels of internal Ca^{2+} determine nerve growth factor dependence of sympathetic ganglion cells. *Proc. natn. Acad. Sci. U.S.A.* **86**, 6421–6425.

Lampe, P.A., Cornbrooks, E.B., Juhasz, A., Franklin, J.L. & Johnson, E.M. Jr 1992 Thapsigargin enhances survival of sympathetic neurons by elevating intracellular calcium concentration ($[Ca^{2+}]_i$). *Soc. Neurosci. Abstr.* **18**, 30.5.

Levi-Montalcini, R. & Booker, B. 1960 Destruction of the sympathetic ganglia in mammals by an antiserum to nerve-growth protein. *Proc. natn. Acad. Sci. U.S.A.* **46**, 384–391.

Martin, D.P., Ito, A., Horigome, K., Lampe, P.A. & Johnson, E.M. Jr 1992 Biochemical characterization of programmed cell death in NGF-deprived sympathetic neurons. *J. Neurobiol.* **23**, 1205–1220.

Martin, D.P., Schmidt, R.E., DiStefano, P.S., Lowry, O.H., Carter, J.G. & Johnson, E.M. Jr 1988 Inhibitors of protein synthesis and RNA synthesis prevent neuronal death caused by nerve growth factor deprivation. *J. Cell Biol.* **106**, 829–844.

Nishi, R. & Berg, D.K. 1981 Effects of high K^+ concentrations on the growth and development of ciliary ganglion neurons in cell culture. *Devl Biol.* **87**, 301–307.

Olson, L. 1993 NGF and the treatment of Alzheimer's disease. *Expl Neurol.* **124**, 5–15.

Oppenheim, R.W. 1991 Cell death during development of the nervous system. *A. Rev. Neurosci.* **14**, 453–501.

Oppenheim, R.W., Prevette, D., Tytell, M. & Homma, S. 1990 Naturally occurring and induced neuronal death in the chick embryo in vivo requires protein and RNA synthesis: evidence for the role of cell death genes. *Devl Biol.* **138**, 104–113.

Pilar, G. & Landmesser, L. 1976 Ultrastructural differences during embryonic cell death in normal and peripherally deprived ciliary ganglia. *J. Cell Biol.* **68**, 339–356.

Putney, J.W. Jr 1993 The signal for capacitative calcium entry. *Cell* **75**, 199–201.

Rich, K.M., Luszczynski, J.R., Osborne, P.A. & Johnson, E.M. Jr 1987 Nerve growth factor protects adult sensory neurons from cell death and atrophy caused by nerve injury. *J. Neurocytol.* **16**, 261–268.

Schmidt, M.F. & Kater, S.B. 1993 Fibroblast growth factors, depolarization, and substratum interact in a combinatorial way to promote neuronal survival. *Devl Biol.* **158**, 228–237.

Scott, B.S. & Fisher, K.C. 1970 Potasssium concentration and number of neurons in cultures of dissociated ganglia. *Expl Neurol.* **27**, 16–22.

Thastrup, O., Cullen, P.J., Drøbak, B.K., Hanley, M.R. & Dawson, A.P. 1990 Thapsigargin, a tumor promoter, discharges intracellular Ca^{2+} stores by specific inhibition of the endoplasmic reticulum Ca^{2+}-ATPase. *Proc. natn. Acad. Sci. U.S.A.* **87**, 2466–2470.

Tolkovsky, A.M., Walker, A.E., Murrell, R.D. & Suidan, H.S. 1990 Ca^{2+} transients are not required as signals for long-term neurite outgrowth from cultured sympathetic neurons. *J. Cell Biol.* **110**, 1295–1306.

Tsao, H., Aletta, J.M. & Greene, L.A. 1990 Nerve growth factor and fibroblast growth factor selectively activate a protein kinase that phosphorylates high molecular weight microtubule-associated proteins. *J. biol. Chem.* **265**, 15471–15480.

Wakade, A.R., Wakade, T.D., Malhotra, R.K. & Bhave, S.V. 1988 Excess K^+ and phorbol ester activate protein kinase C and support the survival of chick sympathetic neurons in culture. *J. Neurochem.* **51**, 975–983.

5

Apoptosis in the haemopoietic system

G. J. COWLING AND T. M. DEXTER

CRC Department of Experimental Haematology, Paterson Institute for Cancer Research, Christie (NHS Trust) Hospital, Manchester M20 9BX, U.K.

SUMMARY

Our previous studies have shown that haemopoietic stem cells undergo apoptotic death as a consequence of growth factor withdrawal. In this paper we review the new data that has accumulated since this observation and compare it with older data from the 'pre-apoptotic' age. Models of erythropoiesis and granulopoiesis that incorporate apoptosis as a normal physiological process controlling homeostasis are examined. The converse to cell death is cell survival, and we describe experiments which suggest that haemopoietic growth factors can not only act as mitogenic or differentiation stimuli but also act as survival signals. We, and others, have proposed that these growth factor-induced survival signals act through the membrane bound polypeptide receptors and share common features of signal transduction with proliferative responses. Enforced expression of *bcl-2* in haemopoietic stem cells is able to overcome apoptosis following the withdrawal of growth factor, and the cells commit into different lineage differentiation programmes. Such cells spontaneously differentiate without cell division, suggesting a stochastic model of haemopoiesis in which the major role of haemopoietic growth factors is to suppress apoptosis and act as mitogens. We review the evidence that the underlying causes of some haematological diseases may be associated with change in the balance between cell survival and death.

1. INTRODUCTION

Blood cells arise from a small number of pluripotent stem cells that are found in the bone marrow. The process is controlled by a range of cytokines which individually bind to their own or shared receptors and thereby modify the proliferation, differentiation and maturation of these cells. These cytokines include interleukin-1 (IL-1) to interleukin-11 (IL-11), erythropoietin (EPO) and the haemopoietic cell colony-stimulating factors (CSFs), stem cell factor (SCF) (also known as *kit* ligand or mast cell growth factor), transforming growth factor β (TGF-β), macrophage inflammatory protein 1α (MIP-1α) and various others, such as interferons and insulin-like growth factors (IGF) (Heyworth *et al.* 1990; Broxmeyer 1992; Metcalf 1993). For many years the emphasis of haemopoietic growth factor research has been on the coupled proliferative and differentiation functions of these proteins, and most molecular studies are aimed at establishing the precise mechanisms by which individual receptors transduce these different signals. However, another underlying, and more subtle, role of growth factors in haemopoietic cell regulation is as survival factors (see table 1).

Since the general acceptance of apoptosis as a mechanism of cell death and the observation that haemopoietic cells undergo apoptotic death (Williams *et al.* 1990) as a consequence of hacmopoietic growth factor redrawal, models of haemopoietic cell regulation have been developed that incorporate this

feature (Koury 1992). Moreover, we have previously suggested that cell death is one of the normal physiological processes that controls homeostasis within the haemopoietic system (Dexter *et al.* 1986).

2. ERYTHROPOIESIS

Erythropoietin (EPO) is a 30–34 kDa glycoprotein produced by the kidney, with smaller amounts made in the foetal and neonatal liver and by resident macrophages in adult tissues. It acts as the primary regulator of mammalian erythropoiesis, stimulating both the proliferation and differentiation of immature blast-forming units-erythroid (BFU-E) and their more-mature progeny, the colony-forming units-erythroid (CFU-E). A third role for EPO is that of a survival factor; it acts on all EPO-responsive cell compartments and withdrawal of growth factor results in apoptotic death (see figure 1). Unlike most of the other haemopoietic growth factors, EPO circulates in the blood, its production being controlled by a feedback mechanism through kidney hypoxia. In circumstances where there is a deficiency in erythrocytes (anaemia) EPO production in the kidney is enhanced, whereas in polycythaemic conditions EPO production is reduced.

Early observations showed that the EPO responsiveness of mice maintained in a polycythaemic condition for more than 40 days by red blood cell transfusions was similar to that seen in normal mice. Furthermore the cycling rate of the EPO responsive

Table 1. *The identified survival and apoptotic characteristics of various haemopoietic growth factors*

(Abbreviations are defined in the text. (?) signifies unknown.)

growth factor	proliferation	survival	apoptotic activity
IL-1	+	+	−
IL-2	+	?	−
IL-3	+	+	−
IL-5	+	+	−
IL-6	+	+	−
GM-CSF	+	+	−
G-CSF	+	+	−
M-CSF	+	?	−
EPO	+	+	−
SCF	+	+	−
MIP-1α	−	?	−
IFN-γ	−	+	−
FAS antigen	?	?	+
TGF-β	−	−	+

cells, as measured by [³H]-thymidine killing, also remained unchanged. These results indicate that continuous production and amplification of EPO-responsive progenitor cells continues for long periods in the absence of mature erythrocyte demand (Lord 1976; Wangenheim *et al.* 1977) and furthermore that the size of this population is not changing. It was noted, at the time, that marrow from such animals showed a high degree of spontaneous cell death (B. I. Lord, personal communication). An anecdotal report of two untreated polycythaemia vera patients also showed substantial numbers of phenotypically normal primitive erythroid progenitor cells which failed to mature (Eaves & Eaves 1978). Such results led to the speculation that EPO was required for not only the

late-stage erythroid progenitor cell development but more importantly their continued survival. This EPO-dependent period extends from at least the CFU-E stage to the stage at which haemoglobin synthesis begins. Koury & Bondurant (1990) demonstrated that the EPO-dependent cells, isolated from the spleens of mice infected with anaemia-inducing strain of Friend leukaemia virus (FVA cells), deprived of EPO, accumulated DNA cleavage fragments characteristic of those found in apoptotic cells by 2–4 h and began dying by 16 h. In the presence of EPO, the progenitor cells survived and developed into reticulocytes. These and other data led Koury (1992) to propose a general model of blood cell production controlled by the suppression of apoptosis by haemopoietic growth factors. Although this 'supply and demand' model (see figure 1) fits most of what we know about EPO-controlled erythropoiesis, it is dependent on a growth factor produced distal to the bone marrow.

Earlier studies of various EPO-responsive progenitor cell populations indicated that, within a mixed population of erythroid progenitor cells, there can be large differences in EPO sensitivities (Eaves & Eaves 1978). These results were supported by EPO binding studies on purified murine CFU-E and their descendants (Landschulz *et al.* 1989). Several observations arose from this study. First, EPO bound on these cells is in rapid flux having a cell surface half-life for [¹²⁵I]-EPO internalization of around 5 min. Second, repeated occupancy of the EPO-receptor (EPOR) is required for mitogenic response. Third, there is an apparent disappearance of high-affinity sites and a persistence of low-affinity sites as the cells mature. The authors suggested that at least two gene products mediate EPO-binding. When the cloned cDNA for

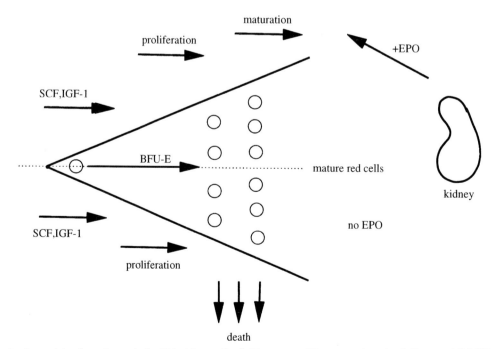

Figure 1. A model of erythropoiesis. Primitive cells (left) can proliferate under the influence of SCF, IGF-1, other conditioning factors and EPO in normal haemopoiesis (top half of diagram), but in absence of EPO (through hypoxic feedback of EPO production or disease), the BFU-Es cannot further proliferate or mature and as a consequence die by apoptosis (bottom half of diagram).

the EPOR (derived from murine erythroleukaemic cells with only low-affinity receptors) is transfected into monkey kidney cells that have no EPO receptors, both high- and low-affinity receptors are found (D'Andrea *et al.* 1989). This observation and other experiments (Dong & Goldwasser 1993) suggest that an unknown accessory protein or co-factor is required for high-affinity EPORs. These high-affinity sites appear to mediate the growth function of EPO on early progenitor cells and the lingering low-affinity receptors have, as yet, no recognized growth or differentiation function. While it is tempting to speculate that such low-affinity receptors may have a role to play in transducing a survival signal, this has yet to be investigated.

In the EPO-dependent cell line HCD-57, EPO can act as both a mitogen and a survival factor, even in cells that are not terminally differentiating (Spival *et al.* 1991). What of other growth factors which allow the survival of erythroid progenitor cells? Using purified murine CFU-E, IGF-1 (when EPO levels are low) also protected the most primitive cells against apoptosis but lost effectiveness with maturity (Boyer *et al.* 1992). IL-3 and SCF alone have no effect in reducing apoptosis in murine foetal liver cells but, when combined with EPO, enhanced activity was seen (Yu *et al.* 1993). In human erythroid progenitors, apoptosis is decreased by SCF and IGF-1 as well as EPO (Muta & Krantz 1993).

3. GRANULOPOIESIS

The dependence of early progenitor cells and mature cells (such as neutrophils and eosinophils) on exogenous growth factors proved to be a sensitive microassay for colony stimulating factors (Begley *et al.* 1986). Furthermore, the observation that the removal of growth factor, at any stage of the haemopoietic development programme leads to a cessation of growth and to death of the developing clone, led to the suggestion that programmed death has a physiological significance in haemopoiesis (Dexter *et al.* 1986). This was later shown experimentally by demonstrating that the withdrawal of the relevant CSF from haemopoietic precursor cell lines such as FDCP-1 and FDCP-Mix, results in active cell death with morphological features and nucleosomal DNA laddering patterns characteristic of apoptosis (Williams *et al.* 1990). Death was not immediate, but took place over a 20–30 h period and viability could be maintained for this period of time by the addition of cycloheximide. Moreover, sublines of FDCP-Mix that were dependent on either IL-3, GM-CSF or G-CSF, behaved in a similar fashion when deprived of growth factor. These data led to the first proposal that apoptosis is a positive control mechanism that regulates haemopoietic precursor cell survival. Many similar studies followed in which both rodent and human haemopoietic cells as well as leukaemic cell lines were also observed to undergo apoptosis in absence of the various growth factors.

That the ability of growth factors to suppress apoptosis is not restricted to normal or leukaemic

cell lines, but also operates on normal early progenitor cell populations, has been shown by several groups. For example, the *c-kit* ligand (stem cell factor (SCF)) has been shown to directly act on highly enriched committed murine progenitor cells in serum-deprived conditions to promote survival, proliferation and development (Heyworth *et al.* 1992); murine mast cells undergo apoptosis on removal of IL-3, an event that is prevented by the addition of SCF, suggesting that these factors act in concert on mast cells (Mekori *et al.* 1993); G-CSF acts as a survival factor in mature neutrophils (Yamamoto *et al.* 1993, Lee *et al.* 1993; Haslett *et al.*, this volume): insulin-like growth factor 1 (IGF-1) has also been shown to delay apoptosis in myeloid precursor cell lines (Rodriguez-Tarduchy *et al.* 1992). These and other data have led us to propose the model outlined in figure 2.

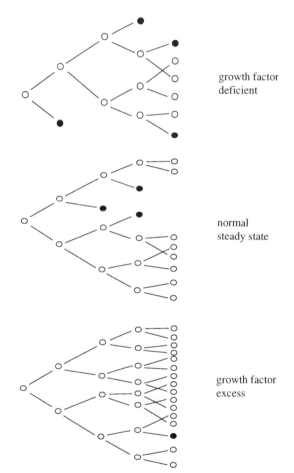

growth factor deficient

normal steady state

growth factor excess

Figure 2. Proposed models of haemopoiesis when growth factor concentration is deficient (top), normal (middle) and in excess (bottom). Open circles are viable cells and closed circles are apoptotic cells. In the primitive compartment of haemopoiesis, stem cells can probably undergo limited self renewal. Withdrawal of the growth factors known to act on the very primitive cell population can lead to apoptosis and therefore such stem cells can die in the absence of growth factor (see top). We speculate that the stroma plays an important role in preventing this happening to any great extent. As stem cells are recruited into the progenitor compartment, they undergo amplification, at each stage (or division) cells can either withdraw from proliferation and differentiate (assuming that the lineage commitment had occurred in the stem cell that they were derived from) or, in the absence of growth factor, die by apoptosis.

Of the protein factors that are known to be growth inhibitory to haemopoietic stem cells, interferon-gamma (IFN-γ) in the presence of either IL-3 or GM-CSF or M-CSF inhibited highly enriched mouse granulocyte-macrophage colony-forming cells (GM-CFC) forming colonies in soft agar with an overall increase in macrophage differentiation (Kan *et al.* 1991). However, in the absence of growth factors, IFN-γ acted as a survival factor and activated the Na^+/H^+ antiport system suggesting that this factor has a dual effect on GM-CFC, decreasing the rate of death but also limiting the proliferative response to CSFs. Members of the IL-8 family of genes including RANTES, MIP-1β and MIP-1α are known to inhibit the growth of haemopoietic cells by a mechanism which involves removal from the cell cycle (Lord *et al.* 1993). The receptor for MIP-1α has been recently reported and its structure suggests that a heterotrimeric G protein may be involved in its signal transduction (Neote *et al.* 1993). Although acting as a haemopoietic cell growth inhibitory factor, it is unclear whether its mechanism involves the suppression of apoptosis.

Transforming growth factor-β1 (TGF-β1) is an ubiquitous cytokine which inhibits the growth of a wide range of cell types including haemopoietic cells *in vitro* (Hino *et al.* 1988; McNiece *et al.* 1992). Although the actions of TGF-β1 on myeloid cells appear complex, with differing responses depending on the cell population, at least some of the effects of TGF-β can be attributed to its ability to induce apoptosis in normal myeloid precursors (Lotem & Sachs 1990) and myeloid leukaemic cell lines (Lotem & Sachs 1992). It is also intriguing that G-CSF, IL-6 and to a lesser extent IL-1 can inhibit this TGF-β induced apoptosis in at least some cells. In addition, varying patterns of response to TGF-β1 were shown by growth-factor-dependent human AML cell lines as well as primary AML cells. These different responses were shown to be due to differences in the TGF-β1-induced apoptosis of AML-target cells (Taetle *et al.* 1993).

4. SURVIVAL VERSUS PROLIFERATION SIGNALS

IL-3 can stimulate the activation of an amiloride-sensitive Na^+/H^+ exchange via protein kinase C activation and it was suggested that the resulting increase in intracellular pH acts as a signal for cellular survival and proliferation in myeloid progenitor cells (Whetton *et al.* 1988; Cook *et al.* 1989; Rodriguez-Tarduchy *et al.* 1990). Recent studies, using IL-3 and GM-CSF cell lines and combinations of PKC activators and inhibitors, have further suggested that the sequential activation of PKC and of the Na^+/H^+ antiporter result in the suppression of apoptosis in target cells (Rajotte *et al.* 1992). Little is known of associated cytoplasmic signalling proteins in these processes but is interesting to note that in one report the inhibition of *c-fes*, a member of the cytoplasmic tyrosine kinase family, by antisense oligomers, was shown to increase the rate of apoptosis in HL60 cells

following chemical induction to granulocyte differentiation (Manfredini *et al.* 1993). Because of the great diversity of growth factors known or thought to act as suppressors of apoptosis and the differing signal transduction pathways each of these factors use, the 'survival signal' is probably a general cellular change. However, although the activation of the Na^+/H^+ antiport system and the subsequent alkalinization provides an attractive mechanism, such changes are also associated with growth-factor-induced proliferation of cells and it is not known what distinguishes a 'proliferative signal' from a 'survival' one. What is clear, however, is that the survival signal can be uncoupled from proliferation, as low concentrations of growth factors such as M-CSF can promote the survival (but not the proliferation) of target cells, while higher concentrations promote survival and proliferation (Stanley & Guilbert 1981; Tushinski *et al.* 1982). To date, detailed biochemical studies that give a molecular definition of this distinction between survival and proliferation signal pathways have yet to be performed.

(a) Survival effectors

Little is known of the effector molecules involved in transduction of signals that suppress apoptosis of haemopoietic cells. The inappropriate expression of certain genes, in cells known to undergo apoptosis when deprived of growth factor, has begun to point to their involvement in growth factor-dependent survival. The induction of *c-myc* gene expression is an immediate early response to mitogenic signals and its expression is decreased in response to growth factor deprivation or growth inhibitors. Although the precise biochemical function of *c-myc* still remains unclear, the downregulation of *c-myc* is a required event for cells to withdraw from the cell cycle and either enter differentiation or a quiescent state. Enforced expression of *c-myc* drives cell cycle progression and blocks differentiation whereas reducing *c-myc* expression (by antisense constructs) blocks progression into S phase and promotes differentiation. In both haemopoietic and other cell systems, it has been shown that failure to down regulate *c-myc* expression in response to IL-3 deprivation prevents cells from undergoing G1 arrest and accelerates apoptosis (Evan *et al.* 1992; Askew *et al.* 1993; Selvakumaran *et al.* 1993). On the other-hand, in some cell systems, HL60 (Beere *et al.* 1993) and lymphoid cell-line CCRF-CEM (Yuh & Thompson 1989) *c-myc* expression is not a prerequisite for apoptosis. Although another nuclear protein, p53, when overexpressed as the wild-type form was reported to induce apoptosis in a range of cell types including haemopoietic cells (Yonish-Rouach *et al.* 1991), this effect was not seen in other cell systems expressing high levels of wild-type p53 (see also Lane, this volume).

The *bcl-2* proto-oncogene has been widely associated with the prevention or delay in apoptosis, in response to different apoptotic stimuli, in a variety of haemopoietic cells. A related gene, *bcl-x*, has been isolated and the protein product of the longer mRNA form can also prevent apoptosis following growth factor withdrawal in IL-3-dependent haemopoietic cells in a similar fashion to *bcl-2* after growth factor withdrawal whereas expression of the shorter messenger RNA product, *bcl-x*$_S$, displays an inhibitory effect on the anti-apoptotic activity of both *bcl-x*$_L$ and *bcl-2* (Boise *et al.* 1993). Expression of *bcl-2* prevents apoptosis in response to IL-3, IL-4 and GM-CSF withdrawal (Vaux *et al.* 1988; Nunez *et al.* 1990). Cell death induced by *c-myc* is also inhibited by *bcl-2* (Bissonnette *et al.* 1992; Fanidi *et al.* 1992).

(b) *Differentiate or die*

Enforced expression of *bcl-2* allows cells to survive in a variety of conditions where they would normally die via apoptosis. We have used the enforced expression of the human *bcl-2* gene to overcome the apoptosis that is seen in the murine myeloid progenitor stem cell line, FDCP-mix, after withdrawal of IL-3 and other cytokines (Williams *et al.* 1990). To our surprise, when progenitor stem cells are unable to either proliferate (no IL-3) or die through apoptosis (high expression of *bcl-2*), we found that the cells instead committed into one of the lineage differentiation programmes, i.e. differentiation by 'default' (Fairbairn *et al.* 1993). By monitoring the fate of single cells with time it was established that proliferation was not required for differentiation and that it was not dependent on the constitutive production of several growth factors or serum components. In other words, one conclusion from these experiments is that the major role for haemopoietic growth factors is to suppress apoptosis and act as mitogens and that they are not required for differentiation. Although such data support the stochastic models of haemopoietic development (Nakahata *et al.* 1982; Suda *et al.* 1984), this does not, however, rule out a role for haemopoietic growth factors in modifying the choice of lineage pathway at an early stage of development. Furthermore our ability to uncouple the processes of proliferation, survival and differentiation should allow further analysis of the molecular mechanisms involved in these processes.

5. APOPTOSIS AND HAEMATOLOGICAL DISEASE

Changes in the balance between cell survival and death have clear implications in underlying causes of some haematological diseases including anaemias such as β-thalassaemia (Yuan *et al.* 1993) and chronic myeloid leukaemia. In the latter, for example, the disease is associated with a chromosome translocation (9;22) that results in the expression of a chimeric *bcr-abl* gene product that has elevated tyrosine kinase activity and which appears to be essential for transformation of cells *in vitro*. Although expression of the cellular *c-abl* gene is not critical for development (Tybulewicz *et al.* 1991), the transfection of murine myeloid stem cells with temperature sensitive mutants of the p210 *bcr-abl* gene (Carlesso *et al.* 1994) and p160 *v-abl* gene (Spooncer *et al.* 1994) has shown that expression of *abl* is associated with a reduced rate of apoptosis at low growth factor levels and an exaggerated proliferative response to low levels of growth factor.

Work from this laboratory also shows that expression of the *v-abl* tyrosine kinase (at permissive temperature) in multipotent myeloid progenitor cells leads to a delay in maturation with a concomitant increase in cell production (Spooncer *et al.* 1994). Thus, a combination of delayed apoptosis and enhanced proliferative ability of cell populations in response to reduced growth factor levels may be the mechanism that provides human CML cells expressing *bcr-abl* with a selective advantage over their normal counterparts (figure 3). Although the biochemical mechanisms for this effect are not known, a general reassessment of the role that apoptosis plays in other haematological diseases is clearly warranted.

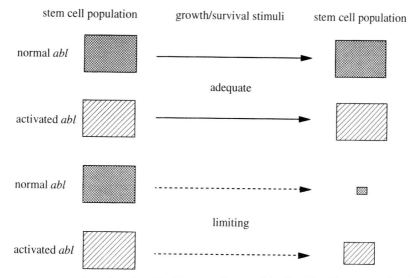

Figure 3. The effect of normal and activated *abl* expression combined with either normal or limited growth factor on the size of the stem cell populations.

REFERENCES

Askew, D.S., Ihle, J.N. & Cleveland, J.L. 1993 Activation of apoptosis associated with enforced *myc* expression in myeloid progenitor cells is dominant to the suppression of apoptosis by interleukin-3 or erythropoietin. *Blood* **82**, 2079–2087.

Beere, H.M., Hickman, J.A., Morimoto, R.I., Parmar, R., Newbould, R. & Waters, C.M. 1993 Changes in HSC70 and *c-myc* in HL-60 cells engaging differentiation or apoptosis. *Molec. Cell. Different.* **1**, 323–343.

Begley, C.G., Lopez, A.F., Nicola, N.A., Warren, D.J., Vadas, M.A., Sauderson,C.J. & Metcalf, D. 1986 Purified colony-stimulating factors enhance the survival of human neutrophils and eosinophils in vitro: a rapid and sensitive microassay for colony-stimulating factors. *Blood* **68**, 162–166.

Bissonnette, R.P., Echeverri, F., Mahboubi, A. & Green, D.R. 1992 Apoptotic cell death induced by c-*myc* is inhibited by *bcl*-2. *Nature, Lond.* **359**, 552–554.

Boise, L.H., Gonzalez-Garcia, M., Postema, C.E. *et al.* 1993 *bcl-x*, a *bcl-2*-related gene that functions as a dominant regulator of apoptotic cell death. *Cell* **74**, 1–20.

Boyer, S.H., Bishop, T.R., Rogers, O.C., Noyes, A.N., Frelin, L.P. & Hobbs, S. 1992 Roles of erythropoietin, insulin-like growth factor 1, and unidentified serum factors in promoting maturation of purified murine erythroid colony-forming units. *Blood* **80**, 2503–2512.

Broxmeyer, H.E. 1992 Suppressor cytokines and regulation of myelopoiesis: biology and possible clinical uses. *Am J. Pediatr. Hemat. Oncol.* **14**, 22–30.

Carlesso, N., Griffin, J.D. & Druker, B.J. 1994 Use of a temperature-sensitive mutant to define the biological effects of the $p210^{bcr-abl}$ tyrosine kinase on proliferation of a factor-dependent murine myeloid cell line. *Oncogene* **9**, 149–156.

Cook, N., Dexter, T.M., Lord, B.I., Cragoe, E.J.J. & Whetton, A.D. 1989 Identification of a common signal associated with cellular proliferation stimulated by four haemopoietic growth factors in a highly enriched population of granulocyte/macrophage colony-forming cells. *EMBO J.* **8**, 2967–2974.

Dexter, T.M., Whetton, A.D. & Heyworth, C.M. 1986 The relevance of protein kinase C activation, glucose transport and ATP generation in the response of haemopoietic cells to growth factors. In *Oncogenes and growth control* (ed. P. Kahn & T. Graf), pp. 163–169. Berlin, Heidelberg: Springer-Verlag.

Dong, Y.J. & Goldwasser E. 1993 Evidence for an accessory component that increases the affinity of the erythropoietin receptor. *Expl Hemat.* **21**, 483–486.

D'Andrea, A.D., Lodish, H.F. & Wong, G.G. 1989 Expression cloning of the murine erythropoietin receptor. *Cell* **57**, 277–285.

Eaves, C.J. & Eaves, A.C. 1978 Erythropoietin dose-response curves for three classes of erythroid progenitors in normal human marrow and in patients with polycythemia vera. *Blood* **52**, 1196–1210.

Evan, G.I., Wyllie, A.H., Gilbert, C.S. *et al.* 1992 Induction of apoptosis in fibroblasts by *c-myc* protein. *Cell* **69**, 119–126.

Fairbairn, L.J., Cowling, G.J., Reipert, B.M. & Dexter, T.M. 1993 Suppression of apoptosis allows differentiation and development of a multipotent hemopoietic cell line in the absence of added growth factors. *Cell* **74**, 823–832.

Fanidi, A., Harrington, E.A. & Evan, G.I. 1992 Cooperative interaction between c-*myc* and *bcl*-2 proto-oncogenes. *Nature, Lond.* **359**, 554–556.

Heyworth, C.M., Vallance, S.J., Whetton, A.D. & Dexter, T.M. 1990 The biochemistry and biology of the myeloid haemopoietic cell growth factors. *J. Cell Sci. Suppl.* **13**, 57–74.

Heyworth, C.M., Whetton, A.D., Nicholls, S., Zsebo, K. & Dexter, T.M. 1992 Stem cell factor directly stimulates the development of enriched granulocyte-macrophage colony-forming cells and promotes the effects of other colony-stimulating factors. *Blood* **80**, 2230–2236.

Hino, M., Tojo, A., Miyazono, K., Urabe, A. & Takaku, F. 1988 Effects of type transforming growth factors on haemopoietic cells. *Br. J. Haematol.* **70**, 143–147.

Kan, O., Heyworth, C.M., Dexter, T.M., Maudsley, P.J., Cook, N., Vallance, S.J. & Whetton, A.D. 1991 Interferon-γ stimulates the survival and influences the development of bipotential granulocyte-macrophage colony-forming cells. *Blood* **78**, 2588–2594.

Koury, M.J. & Bondurant, M.C. 1990 Erythropoietin retards DNA breakdown and prevents programmed death in erythroid progenitor cells. *Science, Wash.* **248**, 378–381.

Koury, M.J. 1992 Programmed cell death (apoptosis) in hematopoieis. *Expl Hemat.* **20**, 391–394.

Landschulz, K.T., Noyes, A.N., Rogers, O. & Boyer, S.H. 1989 Erythropoietin receptors on murine erythroid colony-forming units: natural history. *Blood* **73**, 1476–1486.

Lee, A., Whyte, M.K.B. & Haslett C. 1993 Inhibition of apoptosis and prolongation of neutrophil functional longevity by inflammatory mediators. *J. Leukocyte Biol.* **54**, 283–288.

Lord, B.I. 1976 Stem cell reserve and its control. In *Stem cells of renewing populations* (ed. A. B. Cairnie, P. K. Lada & D. G. Osmond), pp. 165–179. New York, San Francisco, London: Academic Press.

Lord, B.I., Heyworth, C.M. & Woolford, L.B. 1993 Macrophage inflammatory protein: its characteristics, biological properties and role in the regulation of haemopoiesis. *Int. J. Haematol.* **57**, 197–206.

Lotem, J. & Sachs, L. 1990 Selective regulation of the activity of different hematopoietic regulatory proteins by transforming growth factor 1 in normal and leukemic myeloid cells. *Blood* **76**, 1315–1322.

Lotem, J. & Sachs, L. 1992 Hematopoietic cytokines inhibit apoptosis induced by transforming growth factor 1 and cancer chemotherapy compounds in myeloid leukemic cells. *Blood* **80**, 1750–1757.

Manfredini, R., Grande, A., Tagliafico, E. *et al.* 1993 Inhibition of *c-fes* expression by an antisense oligomer causes apoptosis of HL60 cells induced to granulocytic differentiation. *J. exp. Med.* **178**, 381–389.

McNiece, I.K., Bertoncello, I., Keller, J.R., Ruscetti, F.W., Hartley, C.A. & Zsebok, M. 1992 Transforming growth factor inhibits the action of stem cell factor on mouse and human hematopoietic progenitors. *Int. J. Cell Cloning* **10**, 80–86.

Mekori, Y.A., Oh, K.C. & Metcalfe, D.D. 1993 IL-3-dependent murine mast cells undergo apoptosis on removal of IL-3. *J. Immunol.* **151**, 3775–3784.

Neote, K., DiGregorio, D., Mak, J.Y., Horuk, R. & Schall, J.J. 1993 Molecular cloning, functional expression and signalling characteristics of a CC chemokine receptor. *Cell* **72**, 415–425.

Metcalf, D. 1993 Hematopoietic regulators: redundancy or subtlety. *Blood* **82**, 3515–3523.

Muta, K. & Krantz, S.B. 1993 Apoptosis of human erythroid colony-forming cells is decreased by stem cell factor and insulin-like growth factor I as well as erythropoietin. *J. Cell Physiol.* **156**, 264–271.

Nakahata, T., Gros, A.J. & Ogawa, M. 1982 A stochastic

model of self renewal and commitment to differentiation of the primitive hemopoietic stem cells in culture. *J. Cell Physiol.* **113**, 455–458.

Nunez, G., London, L., Hockenbery, D., Alexander, M., McKearn, J. & Korsmeyer, S.J. 1990 Deregulated *Bcl-2* gene expression selectively prolongs survival of growth factor-deprived hematopoietic cell lines. *J. Immunol* **144**, 3602–3610.

Rajotte, D., Haddad, P., Haman, A., Cragoe, E.J. Jr & Hoang T. 1992 Role of protein kinase C and the Na$^+$/H$^+$ antiporter in suppression of apoptosis by granulocyte macrophage colony-stimulating factor and interleukin-3. *J. biol. Chem.*, 267, 9980–9987.

Rodriguez-Tarduchy, G., Collins, M. & Lopez-Rivas, A. 1990 Regulation of apoptosis by interleukin-3-dependent hemopoietic cells by interleukin-3 and calcium ionophores. *EMBO J.* **9**, 2997–3002.

Rodriguez-Tarduchy, G., Collins, M.K.L., Garcia, I. & Lopez-Rivas, A. 1992 Insulin-like growth factor-I inhibits apoptosis in IL-3-dependent hemopoietic cells. *J. Immunol.* **149**, 535–540.

Selvakumaran, M., Liebermann, D. & Hoffman-Liebermann, B. 1993 Myeloblastic leukaemia cells conditionally blocked by myc-estrogen receptor chimeric transgenes for terminal differentiation coupled to growth arrest and apoptosis. *Blood* **81**, 2257–2262.

Spivak, J.L., Pham, T., Isaacs, M. & Hankins, W.D. 1991 Erythropoietin is both a mitogen and a survival factor. *Blood* **77**, 1228–1233.

Spooncer, E., Fairbairn, L.J., Cowling, G.J., Dexter, T.M., Whetton, A.D. & Owen-Lynch, P.J. 1994 Biological consequences of p160^{v-abl} protein tyrosine kinase activity in a primitive multipotent haemopoietic cell line. *Leukemia* **8**, 620–630.

Stanley, E.R. & Guilbert, J. 1981 Methods for the purification assay, characterisation and target cell binding of a colony stimulating factor (CSF-1). *J. Immunol. Meth.* **445**, 253–289.

Suda, T., Suda, J. & Ogawa, M. 1984 Disparate differentiation in mouse hemopoietic colonies derived from paired progenitors. *Proc. natn. Acad. Sci. U.S.A.* **81**, 2520–2524.

Taetle, R., Payne, C., Dos Santos, B., Russel, M. & Segarini, P. 1993 Effects of transforming growth factor β1 on growth and apoptosis of human myelogenous leukemic cells. *Cancer Res.* **53**, 3386–3393.

Tushinski, R.J., Oliver, I.T., Guilbert, L.J., Tynan, P.W., Warner, J.R. and Stanley, E.R. 1982 Survival of mononuclear phagocytes depends on a lineage-specific growth factor that the differentiated cells selectively destroy. *Cell* **28**, 71–81.

Tybulewicz, V.L.J., Crawford, C.C., Jackson, P.K., Bronson, R.T. & Mulligan, R.C. 1991 Neonatal lethality and lymphopenia in mice with a homozygous disruption of the c-*abl* proto-oncogene. *Cell* **65**, 1153–1163.

Vaux, D.L., Cory, S. & Adams, J.M. 1988 *Bcl-2* gene promotes haemopoietic cell survival and cooperates with c-*myc* to immortalise pre-B cells. *Nature, Lond.* **335**, 440–442.

von Wangenheim, H.R., Schofield, R., Kyffin, S. & Klein, B. 1977 Studies on erythroid-committed precursor cells in the polycythaemic mouse. *Biomedicine* **27**, 337–340.

Whetton, A.D., Vallance, S.J., Monk, P.N., Cragoe, E.J., Dexter, T.M. & Heyworth, C.M. 1988 Interleukin-3-stimulated haemopoietic stem cell proliferation. *Biochem. J.* **256**, 585–592.

Williams, G.T., Smith, C.A., Spooncer, E., Dexter, T.M. & Taylor, D.R. 1990 Haemopoietic colony stimulating factors promote cell survival by suppressing apoptosis. *Nature, Lond.* **343**, 76–79.

Yamamoto, C., Yoshida, S., Taniguchi, H., Qin, M.H., Miyamoto, H. & Mizuguchi, Y. 1993 Lipopolysaccharide and granulocyte colony-stimulating factor delay neutrophil apoptosis and ingestion by guinea pig macrophages. *Infect. Immunity* **61**, 1972–1979.

Yonish-Rouach, E., Resnitzky, D., Lotem, J., Sachs, L., Kimchi, A. & Oren, M. 1991 Wild-type p53 induces apoptosis of myeloid leukaemic cells that is inhibited by interleukin-6. *Nature, Lond.* **352**, 345–347.

Yu, H., Bauer, B., Lipke, G.K., Phillips, R.L. & Van Zant, G. 1993 Apoptosis and hematopoiesis in murine fetal liver. *Blood* **81**, 373–384.

Yuan, J., Angelucci, E., Lucarelli, G. *et al.* 1993 Accelerated programmed cell death (apoptosis) in erythroid precursors with severe β-thalassemia. *Blood* **82**, 374–377.

Yuh, Y.S. & Thompson, E.B. 1989 Glucocorticoid effect on oncogene/growth gene expression in human T-lymphoblastic leukaemic cell line CCRF-CEM. *J. biol. Chem.* **264**, 10904–10910.

6

Programmed cell death and the control of cell survival

M. C. RAFF, B. A. BARRES*, J. F. BURNE, H. S. R. COLES, Y. ISHIZAKI AND M. D. JACOBSON

MRC Developmental Neurobiology Programme, MRC Laboratory for Molecular Cell Biology and Biology Department, University College London, London WC1E 6BT, U.K.

SUMMARY

We draw the following tentative conclusions from our studies on programmed cell death (PCD): (i) the amount of normal cell death in mammalian development is still underestimated; (ii) most mammalian cells constitutively express the proteins required to undergo PCD; (iii) the death programme operates by default when a mammalian cell is deprived of signals from other cells; (iv) many normal cell deaths may occur because cells fail to obtain the extracellular signals they need to suppress the death programme; and (v) neither the nucleus nor mitochondrial respiration is required for PCD (or Bcl-2 protection from PCD), raising the possibility that the death programme, like mitosis, is orchestrated by a cytosolic regulator that acts on multiple organelles in parallel.

1. INTRODUCTION

PCD is a fundamental property of animal cells, allowing unwanted cells to be eliminated quickly and neatly (Wyllie *et al.* 1980; Ellis *et al.* 1991). Although the death programme is cell-intrinsic (Wyllie *et al.* 1980; Ellis *et al.* 1991), it is regulated by extracellular signals that can either activate it or suppress it (reviewed in Raff 1992). We have concentrated on signals that suppress PCD and have explored the possibility that most mammalian cells require continuous signalling from other cells to avoid PCD (Raff 1992). Dependence on survival signals would ensure that a cell only survives when and where it is needed, just as dependence on growth factors for proliferation ensures that a cell only divides when a new cell is needed. The importance of such social controls in multicellular organisms is illustrated by the devastating effects of cancer, where the controls are defective.

In this brief review, we first discuss experiments that illustrate the general importance of PCD-suppressing signals and then consider experiments that explore the nature of the death programme itself.

2. PCD IN THE OLIGODENDROCYTE LINEAGE

Oligodendrocytes make myelin in the central nervous system (CNS). Like neurons, they are postmitotic cells that develop from dividing precursors. When either oligodendrocytes or their precursor cells are isolated from the developing rat optic nerve and cultured in the absence of other cell types or exogenous signalling molecules, no matter how high the cell density, they undergo PCD within a day or so (Barres *et al.* 1992). They can be saved by factors released in culture by their normal neighbours (mainly astrocytes) isolated from the optic nerve (Barres *et al.* 1992). They can also be saved by a combination of known growth factors and cytokines. Among the factors that promote the survival of oligodendrocyte lineage cells in culture are platelet-derived growth factor (PDGF), insulin-like growth factor-1 (IGF-1), neurotrophin-3 (NT-3) and ciliary neurotrophic factor (CNTF), all of which are made by astrocytes in culture and are present in the developing optic nerve; whereas a single factor can promote short-term survival in culture, a combination of at least three of these factors is required for long-term survival (Barres *et al.* 1993*b*). Thus neither oligodendrocytes nor their precursors can survive alone in culture: they need signals from other types of cells, and their normal neighbours can provide them; without such signals, the cells kill themselves.

Many oligodendrocytes undergo PCD during normal CNS development. In the developing rat optic nerve, for example, at least half of the oligodendrocytes produced seem to die in this way (Barres *et al.* 1992). The signalling molecules that can promote the survival of oligodendrocytes *in vitro* can also do so *in vivo*: if the levels of PDGF, IGF-1, NT-3 or CNTF are experimentally increased for several days in the developing nerve, the number of dead oligodendrocytes seen in the nerve is greatly reduced and the number of oligodendrocytes is correspondingly increased (Barres *et al.* 1992, 1993*a,b*, 1994). These findings suggest that each of these signalling molecules is normally present in limiting amounts in the developing optic nerve and that many oligodendrocytes die because they fail to receive the signals they

---

* Present address: Department of Neurobiology, Stanford University School of Medicine, Stanford, California 94305-5401, U.S.A.

need to keep their intrinsic death programme suppressed.

What might be the function of the large-scale oligodendrocyte death in the developing optic nerve (and presumably elsewhere in the CNS)? An attractive possibility is that it helps adjust the number of oligodendrocytes to the number (and length) of axons that require myelination (Barres *et al.* 1993*a*), just as normal neuronal death is thought to help adjust the number of developing vertebrate neurons to the number of target cells they innervate (Cowan *et al.* 1984; Barde 1989; Oppenheim 1991). If so, then axons should play an important part in controlling oligodendrocyte survival, and this seems to be the case: if the postnatal optic nerve is cut just behind the eye so that all of the axons in the nerve rapidly degenerate, most of the oligodendrocytes in the nerve selectively die, suggesting that oligodendrocytes normally depend on axons for survival (Barres *et al.* 1993*a*). Although it is unclear whether axons promote oligodendrocyte survival directly or indirectly (by stimulating astrocytes to produce or release survival factors, for example), purified sensory neurons promote the survival of purified oligodendrocytes *in vitro*, suggesting that neurons can act directly, at least in culture (Barres *et al.* 1993*a*).

3. PCD IN THE DEVELOPING KIDNEY

Because cells that undergo PCD in tissues are phagocytosed and degraded quickly and do not induce inflammation, even large-scale normal cell death can be histologically inconspicuous and therefore go unrecognized (Wyllie *et al.* 1980). This was the case for oligodendrocyte death in the developing optic nerve (Barres *et al.* 1992), and it was also the case in the developing kidney. Until recently, cell death was not thought to be a feature of mammalian kidney development, but we found that the proportion of dead cells in frozen sections of developing rat kidney is more than fivefold higher than in the developing optic nerve (Coles *et al.* 1993). We estimate that the amount of normal cell death in the developing kidney may be comparable to that in the developing nervous system, supporting the view that the extent of normal cell death in mammalian development is still greatly underestimated.

If newborn rats are treated systemically with epidermal growth factor (EGF) (Coles *et al.* 1993) or IGF-1 (H. S. R. Coles, unpublished data), the number of dead cells in sections of developing kidney rapidly falls, suggesting that the normal cell death in the kidney, as in the developing nervous system, may reflect the failure of many cells to receive the signals they need to survive. It is possible that many normal cell deaths that occur in other tissues during animal development may also reflect inadequate PCD-suppressing signals.

During kidney development, metanephric mesenchymal cells are induced by invading ureteric bud cells to differentiate into epithelial cells that then form nephrons (Saxen 1987; Ekblom *et al.* 1987; Bard 1992); if, in explant cultures, the mesenchymal cells are deprived of such inducing signals, they undergo PCD, although many can be rescued if EGF is added to the culture medium (Weller *et al.* 1991; Koseki *et al.* 1992). It is therefore possible that many of the cells that die during normal kidney development are mesenchymal cells that either fail to be induced and for this reason fail to respond to new survival signals that are produced and required as development proceeds, or are induced but fail to be included in developing nephrons.

4. PCD IN LENS EPITHELIAL CELLS AND CHONDROCYTES

Although it is clear that some mammalian cells require signals from other cells to avoid PCD, it is not clear that all mammalian cells do. Blastomeres apparently do not: they can survive and cleave in the absence of signals from other cells (Biggers *et al.* 1971). It is possible, however, that once blastomeres differentiate to give rise to two distinct cell types – inner cell mass cells and trophectoderm cells – these cells and all of the nucleated cell types they give rise to become dependent on survival signals from other cells. It is not practical to test this possibility by studying each of the hundreds of mammalian cell types individually. If there are cells that can survive without signals from other cells, however, one might expect lens epithelial cells and cartilage cells (chondrocytes) to be among them, as both lens and cartilage contain only a single cell type and are not vascularized, innervated or drained by lymphatic vessels (Fawcett 1986).

Unlike oligodendrocytes or their precursors, when either lens epithelial cells (Ishizaki *et al.* 1993) or chondrocytes (Ishizaki *et al.* 1994) are cultured at high density in the absence of other cell types or exogenous signalling molecules, they can survive for many weeks, indicating that they do not require signals from other cell types to survive in culture. When cultured at low cell density, however, they undergo PCD. Culture medium from high density cultures promotes the survival of cells in low-density cultures. Thus, lens and cartilage cells seem to require signals from other cells of the same kind to avoid PCD. Such autocrine signalling among cells that reside in tissues composed of a single cell type should perhaps not be surprising.

If lens and cartilage cells need signals from other cells to avoid PCD, it seems likely that most other mammalian cells do also, at least during development and possibly in the adult as well. Chondrocytes isolated from adult rats, for example, require signals from other chondrocytes to survive in culture, just as those isolated from newborn rats do (Ishizaki *et al.* 1994).

5. THE CELL DEATH PROGRAMME

Despite its fundamental importance and apparent evolutionary conservation from nematodes to humans (Vaux *et al.* 1992; Hengartner & Horvitz, this volume), the mechanism of PCD remains unknown. Genetic studies in *C. elegans* have identified two genes, *ced-3* and *ced-4*, that are required for the 131 normal cell deaths that occur during the development of the hermaphrodite (Ellis *et al.* 1991; Hengartner & Horvitz, this volume). The genes have been cloned and sequenced, and *ced-3* has been shown to encode a protease that is structurally homologous to the interleukin-1β-converting enzyme (ICE) (Yuan & Horvitz 1992; Yuan *et al.* 1993), which can induce PCD when overexpressed in fibroblasts (Miura *et al.* 1993). It is still uncertain, however, how any of these proteins induce cell death; it is unclear, for example, whether they are effectors or activators of the death programme. A third gene, *ced-9*, which acts as a brake on the death programme in *C. elegans* (Hengartner *et al.* 1992), is structurally (Hengartner & Horvitz, this volume) and functionally (Vaux *et al.* 1992; Hengartner & Horvitz, this volume) homologous to the mammalian gene *bcl-2*, which suppresses PCD in many, but not all, mammalian cell types (Vaux *et al.* 1988; Korsmeyer *et al.* 1994). Bcl-2 is now known to be a member of a family of related proteins that regulate PCD in mammalian cells (Boise *et al.* 1993; Oltvai *et al.* 1993).

To test whether most mammalian cells are capable of undergoing PCD, we have used the broad-spectrum protein-kinase inhibitor staurosporine. We reasoned that, if most cells need signals from other cells to avoid PCD, inhibition of many of the protein kinases involved in the intracellular signalling pathways activated by survival signals should induce most cells to undergo PCD. We have found that high concentrations ($\geqslant 1 \mu M$) of staurosporine induce PCD in all of the many mammalian cell types that we have tested; the only exception so far is mouse blastomeres (Jacobson *et al.* 1993; Ishizaki *et al.* 1993; Jacobson *et al.* 1994; H. S. R. Coles, K. Raff, M. C. Raff, T. J. Davies & R. L. Gardner, in preparation). Moreover, in all cases tested, protein synthesis inhibitors fail to block, and usually enhance, staurosporine-induced PCD, suggesting that most cells are not only capable of undergoing PCD but constitutively express all of the protein components required to effect the programme. In those cases where RNA or protein synthesis has been shown to be required for the induction of PCD, macromolecular synthesis seems to be required to activate or de-repress the programme, rather than effect it, as the same cells can usually be induced in other ways to undergo PCD in the absence of macromolecular synthesis (see Martin 1993).

Nuclear changes are a prominent feature of PCD, and it has often been suggested that they are the cause of death: degradation of nuclear DNA by endonucleases, for example, may kill the cell. We have found, however, in the two very different cell lines we have tested, that we can remove the nucleus from a cell and the cell will still undergo the characteristic cytoplasmic changes of apoptosis when either treated with staurosporine or deprived of survival signals; the sequence and timing of the changes are indistinguishable in the cytoplast and its nucleated parent cell (Jacobson *et al.* 1994). Moreover, if cytoplasts are prepared from cells that overexpress the Bcl-2 protein, they are protected from PCD (Jacobson *et al.* 1994). These findings suggest that the nucleus is not required for either PCD or Bcl-2 protection.

The Bcl-2 protein was initially thought to be associated with the inner mitochondrial membrane (Hockenbery *et al.* 1990), raising the possibility that the mitochondrion may be the locus of Bcl-2 action and the primary target in PCD. To explore this possibility we studied human fibroblast cell lines that, as a result of prolonged treatment with ethidium bromide, do not have mitochondrial DNA (King & Attardi 1989). These cells have nonfunctional electron transport chains and therefore cannot carry out mitochondrial respiration. Nonetheless, they are just as sensitive as their normal parent cells to staurosporine-induced PCD, or to survival-factor-deprivation-induced PCD, or to the protective effects of Bcl-2, suggesting that mitochondrial respiration is not required for either PCD or Bcl-2 protection (Jacobson *et al.* 1993).

These findings have lead us to suggest that PCD is orchestrated by a cytosolic regulator that acts on multiple organelles in parallel (Jacobson *et al.* 1994), much as the cytosolic regulator M-phase-promoting factor (MPF) orchestrates the mitotic phase of the cell cycle (Nurse 1990). In fact, PCD shares a number of features with mitosis: although the final outcome is very different, in both cases the cells round up, the plasma membrane blebs, the nuclear lamina disassembles, and the chromatin condenses. It has therefore been suggested that PCD may be an abnormal or mistimed mitosis (Ucker 1991; Rubin *et al.* 1993). It seems to us, however, that PCD is too fundamental and important to be an aberrant mitosis. We prefer the view that it is a highly specialized process that may possibly have evolved from the process of mitosis and may share some components with it.

REFERENCES

Bard, J.B.L. 1992 The development of the mouse kidney – embryogenesis writ small. *Curr. Opin. Genet. Dev.* **2**, 589–595.

Barde, Y.A. 1989 Trophic factors and neuronal survival. *Neuron* **2**, 1525–1534.

Barres, B.A., Hart, I.K., Coles, H.C., Burne, J.F., Voyvodic, J.T., Richardson, W.D. *et al.* 1992 Cell death and control of cell survival in the oligodendrocyte lineage. *Cell* **70**, 31–46.

Barres, B.A., Jacobson, M.D., Schmid, R., Sendtner, M. & Raff, M.C. 1993a Does oligodendrocyte survival depend on axons? *Curr. Biol.* **3**, 489–497.

Barres, B.A., Schmid, R., Sendtner, M. & Raff, M.C. 1993b Multiple extracellular signals are required for long-term oligodendrocyte survival. *Development* **118**, 283–295.

Barres, B.A., Raff, M.C., Gaese, F., Bartke, I. & Barde, Y.A. 1994 A crucial role for NT-3 in oligodendrocyte development. *Nature, Lond.* **367**, 371–375.

Biggers, J.D., Whitten, W.K., Whittingham, D.G. 1971 The culture of mouse embryos in vitro. In *Methods in mammalian embryology* (ed. J. C. Daniel), pp. 86–116. San Francisco: Freeman.

Boise, L.H., González-García, M., Postema, C.E., Ding, L., Lindsten, T., Turka, L.A. *et al.* 1993 *bcl-x*, a *bcl-2*-related gene that functions as a dominant regulator of apoptotic cell death. *Cell* **74**, 597–608.

Coles, H.S.R., Burne, J.F. & Raff, M.C. 1993 Large-scale normal cell death in the developing rat kidney and its reduction by epidermal growth factor. *Development* **118**, 777–784.

Cowan, W.M., Fawcett, J.W., O'Leary, D.D.M. & Stanfield, B.B. 1984 Regressive events in neurogenesis. *Science, Wash.* **225**, 1258–1265.

Ekblom, P., Aufderheide, E., Klein, G., Kurz, A. & Weller, A. 1987 Cell interactions during kidney development. In *Mesenchymal-epithelial interactions in neural development* (ed. J. R. Wolff *et al.*), pp. 101–110. Heidelberg: Springer-Verlag.

Ellis, R.E., Yuan, J.Y. & Horvitz, H.R. 1991 Mechanisms and functions of cell death. *A. Rev. Cell Biol.* **7**, 663–698.

Fawcett, D.W. 1986 *Bloom and Fawcett: a textbook of histology*, 11th edn. Philadelphia: W. B. Saunders.

Hengartner, M.O., Ellis, R.E. & Horvitz, H.R. 1992 *Caenorhabditis elegans* gene ced-9 protects cells from programmed cell death. *Nature, Lond.* **356**, 494–499.

Hengartner, M.O. & Horvitz, H.R. 1994 *C. elegans* cell death gene *ced-9* encodes a functional homolog of mammalian proto-oncogene *bcl-2*. *Cell* **76**, 665–676.

Hockenbery, D.M., Nuñez, G., Milliman, C., Schreiber, R.O. & Korsmeyer, S.J. 1990 BCL-2 is an inner mitochondrial membrane protein that blocks programmed cell death. *Nature, Lond.* **348**, 334–336.

Ishizaki, Y., Voyvodic, J.T., Burne, J.F. & Raff, M.C. 1993 Control of lens epithelial cell survival. *J. Cell Biol.* **121**, 899–908.

Ishizaki, Y., Burne, J.F. & Raff, M.C. 1994 Autocrine dependence of chondrocyte survival in culture. *J. Cell Biol.* (In the press.)

Jacobson, M.D., Burne, J.F., King, M.P., Miyashita, T., Reed, J.C. & Raff, M.C. 1993 Bcl-2 blocks apoptosis in cells lacking mitochondrial DNA. *Nature, Lond.* **361**, 365–369.

Jacobson, M.D., Burne, J.F. & Raff, M.C. 1994 Programmed cell death and Bcl-2 protection in the absence of a nucleus. *EMBO J.* **13 13**, 1899–1910.

King, M.P. & Attardi, G. 1989 Human cells lacking mtDNA: repopulation with exogenous mitochondria by complementation. *Science, Wash.* **246**, 500–503.

Korsmeyer, S.J., Shutter, J.R., Veis, D.J., Merry, D.E. & Oltvai, Z.N. 1994 Bcl-2/Bax: a rheostat that regulates an anti-oxidant pathway and cell death. *Semin. Cancer Biol.* (In the press.)

Koseki, C., Herzlinger, D. & Al-Awqati, Q. 1992 Apoptosis in metanephric development. *J. Cell Biol.* **119**, 1327–1333.

Martin, S.J. 1993 Apoptosis: suicide, execution or murder? *Trends Cell Biol.* **3**, 141–144.

Miura, M., Zhu, H., Rotello, R., Hartwieg, E.A. & Yuan, J. 1993 Induction of apoptosis in fibroblasts by IL-1 beta-converting enzyme, a mammalian homolog of the *C. elegans* cell death gene *ced-3*. *Cell* **75**, 653–660.

Nurse, P. 1990 Universal control mechanism regulating onset of M phase. *Nature, Lond.* **344**, 503–508.

Oltvai, Z.N., Milliman, C.L. & Korsmeyer, S.J. 1993 Bcl-2 heterodimerizes in vivo with a conserved homolog, bax, that accelerates programed cell death. *Cell* **74**, 609–619.

Oppenheim, R.W. 1991 Cell death during development of the nervous system. *A. Rev. Neurosci.* **14**, 453–501.

Raff, M.C. 1992 Social controls on cell survival and cell death. *Nature, Lond.* **356**, 397–400.

Rubin, L.L., Philpott, K.L. & Brooks, S.F. 1993 The cell cycle and cell death. *Curr. Biol.* **3**, 391–394.

Ucker, D.S. 1991 Death by suicide: one way to go in mammalian cellular development? *New Biol.* **3**, 103–109.

Vaux, D.L., Cory, S. & Adams, J.M. 1988 Bcl-2 gene promotes haemopoietic cell survival and cooperates with c-*myc* to immortalize pre-B cells. *Nature, Lond.* **335**, 440–442.

Vaux, D.L., Weissman, I.L. & Kim, S.K. 1992 Prevention of programmed cell death in *Caenorhabditis elegans* by human *bcl-2*. *Science, Wash.* **258**, 1955–1957.

Weller, A., Sorokin, L., Illgen, E.-M. & Ekblom, P. 1991 Development and growth of mouse embryonic kidney in organ culture and modulation of development by soluble growth factor. *Devl Biol.* **144**, 248–261.

Wyllie, A.H., Kerr, J.F.R. & Currie, A.R. 1980 Cell death: the significance of apoptosis. *Int. Rev. Cytol.* **68**, 251–306.

Yuan, J. & Horvitz, H.R. 1992 The *Caenorhabditis elegans* cell death gene *ced-4* encodes a novel protein and is expressed during the period of extensive programmed cell death. *Development* **116**, 309–320.

Yuan, J., Shaham, S., Ledoux, S., Ellis, H.M. & Horvitz, H.R. 1993 The *C. elegans* cell death gene *ced-3* encodes a protein similar to mammalian interleukin-1 beta-converting enzyme. *Cell* **75**, 641–652.

7

Integrated control of cell proliferation and cell death by the c-*myc* oncogene

GERARD EVAN, ELIZABETH HARRINGTON, ABDALLAH FANIDI, HARTMUT LAND, BRUNO AMATI AND MARTIN BENNETT

Biochemistry of the Cell Nucleus Laboratory, Imperial Cancer Research Fund, PO Box 123, 44 Lincoln's Inn Fields, London WC2A 3PX, U.K.

SUMMARY

Regulation of multicellular architecture involves a dynamic equilibrium between cell proliferation, differentiation with consequent growth arrest, and cell death. Apoptosis is one particular form of active cell death that is extremely rapid and characterized by auto-destruction of chromatin, cellular blebbing and condensation, and vesicularization of internal components.

The c-*myc* proto-oncogene encodes an essential component of the cell's proliferative machinery and its deregulated expression is implicated in most neoplasms. Intriguingly, c-*myc* can also act as a potent inducer of apoptosis. Myc-induced apoptosis occurs only in cells deprived of growth factors or forcibly arrested with cytostatic drugs. Myc-induced apoptosis is dependent upon the level at which it is expressed and deletion mapping shows that regions of c-Myc required for apoptosis overlap with regions necessary for co-transformation, autoregulation, inhibition of differentiation, transcriptional activation and sequence-specific DNA binding. Moreover, induction of apoptosis by c-Myc requires association with c-Myc's heterologous partner, Max. All of this strongly implies that c-Myc drives apoptosis through a transcriptional mechanism: presumably by modulation of target genes.

Two simple models can be invoked to explain the induction of apoptosis by c-Myc. One holds that death arises from a conflict in growth signals which is generated by the inappropriate or unscheduled expression of c-Myc under conditions that would normally promote growth arrest. In this 'Conflict' model, induction of apoptosis is not a normal function of c-Myc but a pathological manifestation of its deregulation. It thus has significance only for models of carcinogenic progression in which *myc* genes are invariably disrupted. The other model holds that induction of apoptosis is a normal obligate function of c-Myc which is modulated by specific survival factors. Thus, every cell that enters the cycle invokes an obligate abort suicide pathway which must be continuously suppressed by signals from the immediate cellular environment for the proliferating cell to survive. Evidence will be presented supporting this second 'Dual Signal' model for cell growth and survival, and its widespread implications will be discussed.

1. INTRODUCTION

In any population of unicellular organisms, mutants that acquire a proliferative advantage spontaneously overgrow their less vigorous siblings. One of the deepest and most abiding paradoxes of multi-cellularity is how such spontaneous outgrowth of faster-proliferating variants is suppressed while at the same time permitting substantial proliferation of component cells. In man, this problem is even more acute for three reasons: our substantial physical size, our longevity and the self-renewing nature of our tissues. Clearly, the larger an organism becomes, the greater the number of potential cellular targets for neoplastic mutations. Likewise, the longer an organism lives, the greater the chances of neoplasia occurring at some point in life. Finally, many of our tissues (notably epithelial and haematopoietic)

undergo substantial proliferation throughout our lives, again increasing the possibility of neoplastic mutations' occurring during our lives. However, neoplasia is tightly suppressed in metazoans. We know this because, although cancer affects one in three individuals during their lives, cancer is a clonal disease that arises through expansion of a single affected cell. Thus, the successful cancer cell only ever arises in one in three persons, out of all the billions of proliferating cells within our bodies. The cancer cell is, therefore, extremely rare. This rarity is surprising because, in principal, any mutant cell that achieves some growth advantage over its fellows might be expected to undergo clonal expansion and thereby provide an increased target site for yet further carcinogenic mutations. Carcinogenic progression thus appears to be an inevitable consequence of natural selection within the soma, given enough mutations. The extreme rarity of the cancer in man must, therefore, imply the existence of powerful mechanisms to suppress neoplasia. Presumably, this is either by ensuring that rate of mutation in human cells is at an extremely low level or by the existence of mechanisms that eliminate potential tumour cells when they do arise. In this review, we suggest that an intrinsic anti-neoplastic mechanism exists in all cells and which works through the obligatory coupling of cell proliferative and cell suicide pathways.

2. MYC

c-*myc* encodes a short-lived sequence-specific DNA-binding protein whose expression is elevated or deregulated in virtually all tested tumours (Spencer & Groudine 1991). c-*myc* is one of several related *myc* genes present within the mammalian genome (DePinho *et al.* 1991). However, of this family only c-*myc* is expressed in fully differentiated 'adult' cells that retain proliferative capacity (e.g. epithelial, mesenchymal or lymphoid cells). The c-*myc* protein, c-Myc, is most probably a transcription factor. It possesses an N-terminal domain with a transcriptional modulatory domain that engages the basal transcription machinery activity (Kato *et al.* 1990; Amati *et al.* 1992; Kretzner *et al.* 1992) and may also interact with cell cycle regulatory components such as p107 (Gu *et al.* 1994), and a C-terminal DNA-binding/dimerization domain, akin to that present in many other transcription factors of the bHLHZ class, that mediates dimerization with the bHLHZ protein Max (reviewed in Evan & Littlewood 1993). However, to date, few c-Myc target genes have been defined and those that have offer little clue as to the biological function of c-Myc.

Expression of c-*myc* (or of another member of the *myc* gene family) appears necessary and, in some cases, sufficient for cell proliferation. Inhibition of c-*myc* expression with antisense c-*myc* oligonucleotides effectively blocks cell proliferation (Heikkila *et al.* 1987; Loke *et al.* 1988; Prochownik *et al.* 1988; Wickstrom *et al.* 1989; Bennett *et al.* 1993) and ectopic expression of c-*myc* is sufficient to drive quiescent fibroblasts into cycle (Eilers *et al.* 1989) and keep them there, even in the absence of mitogens or the presence of the anti-proliferative cytokine γ-interferon (Evan *et al.* 1992). Thus, c-Myc locks cells in a continuously proliferating state. Regions of the c-Myc protein required for both of these mitogenic effects are identical to those required for c-Myc to act as transcription factor (Evan *et al.* 1992). Thus, c-Myc most probably acts by regulating target genes that control entry into and exit from the cell cycle.

In vitro, deregulation of c-*myc* expression appears sufficient to generate continuously proliferating cells that can no longer respond to cues that would normally trigger their growth arrest, i.e. *de facto* tumour cells. This suggests that deregulation of a *single* gene, c-*myc*, is sufficient to convert a normal cell into tumour cell. However, this conclusion flies in the face of overwhelming evidence that neoplastic conversion is protracted and requires multiple mutations; one

obvious manifestation of this being the extreme rarity of cancer already alluded to above. Thus, even though cells with deregulated c-*myc* expression exhibit unrestrained and unrestrainable proliferation *in vitro*, this cannot be sufficient for neoplastic transformation.

An explanation for why it is that cells with deregulated c-*myc* expression are not fully neoplastic comes from inspection of the growth rate of such cells under conditions where growth factors are limiting. In normal fibroblasts, c-*myc* expression is tightly dependent upon mitogen availability: in the absence of growth factors (i.e. low serum) c-*myc* is rapidly down-regulated and the cells arrest in G1 (Dean *et al.* 1986; Waters *et al.* 1991). However, fibroblasts with deregulated c-*myc* expression continue to cycle in the absence of growth factors but their numbers do not necessarily increase because of substantial cell death. This cell death has all the features of apoptosis (figure 1); it is rapid (20–40 min), accompanied by cell surface blebbing, cell shrinkage and fragmentation, and cell DNA is cleaved into fragments of nucleosome length (Evan *et al.* 1992). Induction of apoptosis by c-Myc is greatest in cells expressing high levels of the protein, but is nonetheless clearly evident in cells expressing levels of c-Myc present in untransformed proliferating fibroblasts. Analogous Myc-dependent apoptosis has been reported also in growth factor-deprived haematopoietic cells (Askew *et al.* 1991), suggesting that induction of apoptosis by c-Myc may be a general phenomenon.

To determine by what molecular mechanism c-Myc triggers apoptosis in serum-deprived fibroblasts, we carried out site-directed mutagenesis on the c-Myc protein. We observed complete overlap of those regions required for c-Myc to function as an oncoprotein (promoting cell proliferation) (Stone *et al.* 1987) and those required to trigger apoptosis in low serum (Evan *et al.* 1992) (figure 2). The regions were the N-terminal transactivation domain, the sequence-specific DNA binding basic regions and the intact C-terminal helix-loop-helix-leucine zipper dimerization domain. The exact same regions are also required for c-Myc to function as a transcription factor. Thus, the activity of c-Myc as an inducer of apoptosis is genetically inseparable from its ability to promote the entirely contradictory function of cell growth, and both functions probably involve the specific modulation of c-Myc of target genes. All known transcription factors of the bHLHZ class require dimerization with a partner in order to bind DNA and exert their action. Therefore, to confirm further that c-Myc controls both mitogenic and apoptotic functions of c-Myc by a transcriptional mechanism, we investigated whether both of these functions required c-Myc to interact with its partner bHLHZ protein Max. Unfortunately, Max is ubiquitously expressed, making it impossible to ask directly if c-Myc can act in the absence of Max. We therefore made use of mutants of c-Myc and Max that have reciprocally exchanged dimerization specificities, such that the c-Myc mutant (MycEG) can no longer interact with wild-type Max but can dimerize with

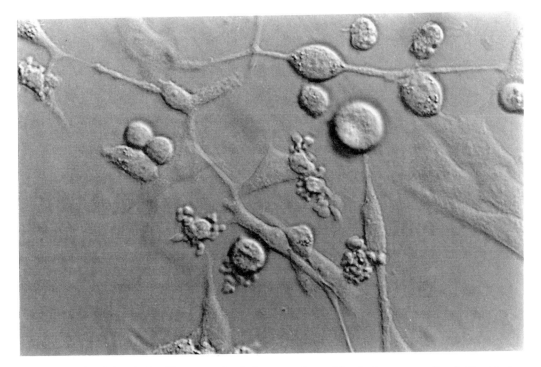

Figure 1. Apoptosis of Rat-1 fibroblasts constitutively expressing c-Myc in low serum. Rat-1 fibroblasts expressing human c-*myc* driven from a retrovirus LTR promoter were cultured in 0.5% foetal calf serum for 24 h. Cells were fixed in 3.5% paraformaldehyde in PBS and visualized by Nomarski Interference Optics.

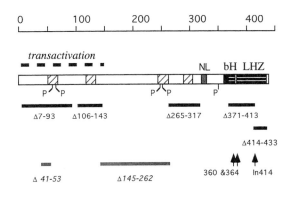

■ Mutants inactive in co-transformation, suppression of growth arrest and induction of apoptosis

▨ Mutants still active in co-transformation, suppression of growth arrest and induction of apoptosis

Figure 2. Mutagenesis study of the human c-Myc protein. Site directed mutants of human c-Myc were expressed in Rat-1 or Swiss 3T3 mouse fibroblasts and the cells assayed for co-transformation in association with activated H-*ras*, inability to arrest growth in low serum and for initiation of apoptosis in low serum. Δ mutants are deletions between the amino acid residues shown, *In* mutants are insertional mutations and the others are point mutations that disrupt the DNA binding domain. Regions of the c-Myc protein marked are the transactivation domain (N-terminal), known phosphorylation sites (P), the dominant nuclear localization signal (NL), the basic DNA-binding region (b) and the C-terminal helix-loop-helix-leucine zipper (HLHZ) dimerization domain.

the corresponding MaxEG mutant (Amati *et al.* 1993). By this strategy we have demonstrated that Max is absolutely required for induction of both cell proliferation and apoptosis by c-Myc. This finding significantly reinforces the notion that both proliferative and apoptotic functions of c-Myc are transcriptional.

3. THE BIOLOGICAL SIGNIFICANCE OF c-MYC-INDUCED APOPTOSIS

The observation that deregulated c-Myc expression triggers apoptosis marries well with a large amount of data indicating that genetic lesions that suppress apoptosis can synergize with c-Myc oncogenically. Transgenic animals whose lymphocytes express deregulated c-*myc* exhibit increased sensitivity to induction of apoptosis in lymphoid organs (Dyall & Cory 1988; Langdon *et al.* 1988; Neiman *et al.* 1991). The anti-apoptotic oncogene *bcl*-2 suppresses such c-Myc-induced apoptosis and synergizes with c-Myc to promote development of lymphomas (Vaux *et al.* 1988). Bcl-2 protein expression also blocks c-Myc induced apoptosis in fibroblasts in response to serum deprivation (Bissonnette *et al.* 1992; Fanidi *et al.* 1992; Wagner *et al.* 1993) or DNA damage (Fanidi *et al.* 1992). More generally, deregulation of c-*myc* is virtually ubiquitous in tumour cells, implying strong selection for c-*myc* activation during carcinogenesis. However, the observation that c-*myc* deregulation is also a potent trigger of apoptosis strongly suggests that anti-apoptotic lesions are likely to be common components of carcinogenesis.

Deregulated expression of c-*myc* is, therefore, a potent trigger of apoptosis. But is there any evidence to suggest that normal c-*myc* expression might be involved in promoting apoptosis? Two simple models can be invoked to explain the induction of apoptosis by c-Myc following serum-deprivation (figure 3). The first (figure 3*a*) argues that cell death arises because of a conflict in signals between the growth promoting action of c-Myc and the growth inhibitory effect of growth factor deprivation. In this model, induction of apoptosis is a pathological consequence of 'inappropriate' c-Myc expression and is not a normal function of c-Myc but arises from a 'conflict of signals'. An alternative, if unorthodox, model for c-Myc-induced apoptosis (figure 3*b*) is to propose that induction of an apoptotic programme is a *bona fide* and obligate component of c-Myc action that necessarily accompanies proliferation, i.e. that proliferation and apoptosis are obligatorily coupled. In this model, successful (i.e. viable) cell proliferation requires two independent signals, one to trigger mitogenesis and the other to suppress the concomitant apoptotic programme. According to this 'Dual Signal' model,

cells expressing c-Myc die in low serum not because of conflict in growth signals but because they are deprived of serum factors required to suppress the c-Myc-induced apoptotic programme. Induction of apoptosis by c-Myc is therefore a normal physiological aspect of c-Myc function.

As outlined above, c-Myc induces apoptosis almost certainly by a transcriptional mechanism. We reasoned that in the 'conflict' model c-Myc only induces its apoptotic transcriptional programme as a result of a conflict of signals. Accordingly, we sought to test the conflict model by blocking execution of the apoptotic programme with either cycloheximide or actinomycin D prior to establishing a conflict of signals (i.e. serum deprivation) (figure 3). Both cycloheximide or actinomycin D block dexamethasone-induced apoptosis in thymocytes which is therefore thought to require *de novo* protein synthesis. Surprisingly, we observed no inhibition of c-Myc-induced apoptosis by either cycloheximide or actinomycin D: indeed, merely adding either drug to otherwise fully viable cells expressing c-Myc in high serum triggered rapid apoptosis. The higher the level of c-Myc expression at the time either drug was added the more rapid and extreme was the apoptosis that occurred. The only conclusion possible from this experiment is that the c-Myc-induced apoptotic transcriptional programme already pre-existed, albeit in silent form, within these cells.

We conclude the following:

1. Induction of apoptosis by c-Myc involves a transcriptional programme (either activation of a pro-apoptotic programme or suppression of an anti-apoptotic programme).

2. The programme pre-exists in viable cells as a consequence of c-Myc expression but is silent in high serum.

3. The c-Myc-induced apoptotic programme becomes active in low serum.

Thus, it seems most likely that specific factors in serum act to suppress the activation of an apoptotic programme that is put in place by c-Myc. This idea is most consistent with the 'Dual Signal' model outlined above in which specific signals regulate the activity of an underlying c-Myc-induced apoptotic programme.

4. CYTOKINES THAT REGULATE c-MYC-INDUCED APOPTOSIS

Replacement of serum with a cytokine-free serum substitute which provides the same essential nutrients as serum does not suppress c-Myc-induced apoptosis in rat fibroblasts, indicating that cell death in low serum does not result from nutritional privation. Next, various cytokines were tested for their abilities to suppress c-Myc-induced apoptosis in serum-deprived cells. Addition of any one of the cytokines IGF-I, IGF-II, insulin, PDGF AB or PDGF BB significantly suppressed apoptosis in the absence of any other exogenous cytokines or nutrients, whereas EGF, basic FGF, acidic FGF, Interleukin-1, TGFα, TGFβ and bombesin all failed to inhibit apoptosis (figure 4).

(*a*)

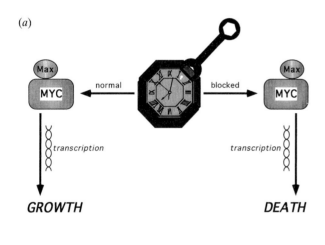

(*b*)

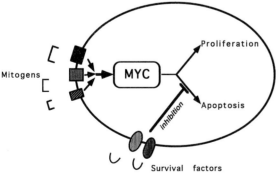

Figure 3. Alternative models to explain c-Myc-induced apoptosis in low serum. (*a*) Death arises because of a conflict of signals. Normally, c-Myc promotes a transcriptional programme leading to cell proliferation. However, in the absence of serum, this programme is interrupted and the resulting conflict directs c-Myc to activate an apoptotic programme. (*b*) 'Dual Signal' hypothesis. Induction of both proliferation and apoptosis are normal consequences of c-Myc expression. Cell fate is determined by ectopic factors that specifically modulate the apoptotic programme.

Inability to suppress apoptosis is not due to lack of appropriate receptors and cognate downstream signal transduction pathways exist for each cytokine, as evidenced by the fact that all cytokines induced transient expression of the immediate early nuclear proteins c-Fos and Egr-1/Zif268/NGFIA. Moreover, EGF, bFGF and bombesin are all potent mitogens for fibroblasts, whereas IGF-I is only poorly mitogenic. Thus, there is no direct correlation between mitogenicity and ability to suppress c-Myc-induced apoptosis among tested cytokines. The anti-apoptotic effects of IGFs and PDGF are evident in all tested cells of mesenchymal origin: primary and immortalized fibroblasts of rodent or human origin and primary rat vascular smooth muscle cells. However, preliminary experiments indicate that other cytokines may fulfil analogous anti-apoptotic roles in other cell lineages.

In the 'conflict of growth signals' model, serum deprivation triggers apoptosis in cells expressing c-Myc because c-Myc alone is insufficient to promote a 'balanced' or 'integrated' mitogenic programme. Consequently, the cells enter a premature or 'unscheduled' cell cycle which results in their death. In this scenario, IGFs and PDGF block c-Myc-induced apoptosis by providing necessary auxiliary growth signals needed to integrate correctly the cell's growth programme. The abilities of IGF-I and PDGF to suppress c-Myc-induced apoptosis is therefore critically dependent upon the activity of both cytokines as mitogens promoting cell cycle progression. In contrast, the 'Dual Signal' hypothesis postulates that IGFs prevent cell death by directly modulating the cell death pathway, irrespective of growth status of the cell. The 'Dual Signal' model therefore makes a unique prediction that the abilities of survival factors to inhibit apoptosis will not be dependent upon the competence of those factors to promote cell cycle progression. To discriminate between the 'Conflict' and 'Dual Signal' hypotheses we therefore asked two key questions. First, can IGF-I suppress c-Myc-induced apoptosis under conditions where it cannot promote cell cycle progression; for example, when cells are profoundly blocked in cycle with cytostatic drugs. Second, does IGF-I suppress c-Myc-induced apoptosis under conditions where it is not required for cell cycle progression. To do this, we examined the effects of IGF-I in post-commitment S/G2 fibroblasts which are factor-independent for completion of their cell cycles.

Cells profoundly blocked in post-commitment parts of the cell cycle by either thymidine (S) or etoposide (S/G$_2$) undergo apoptosis upon activation of c-Myc (Evan *et al.* 1992; Fanidi *et al.* 1992; Harrington *et al.* 1994*a*). Neither IGF-I nor PDGF is able to relieve these potent blocks to cell cycle progression. We therefore asked whether IGF-I exerted any effect on apoptosis induced by either drug in fibroblasts expressing c-Myc. We observed both IGF-I and PDGF to exert significant anti-apoptotic activity in Myc-fibroblasts exposed to thymidine or etoposide. Thus, both cytokines can suppress apoptosis under

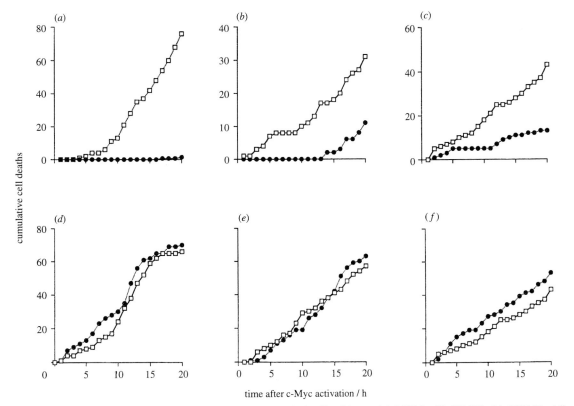

Figure 4. Effects of cytokines on c-Myc-induced apoptosis in fibroblasts. (*a*) IGF-I; (*b*) PDGF; (*c*) IGF-II; (*d*) EGF; (*e*) bFGF; and (*f*) bombesin. One hundred Rat-1 fibroblasts constitutively expressing c-*myc* were cultured in low serum for 20 h in the presence (open squares) or absence (filled circles) of the labelled cytokines (Harrington *et al.* 1994). Apoptotic cell deaths were monitored by time-lapse videomicroscopy. Cumulative cell deaths were plotted against time in hours.

conditions where they are unable to promote cell proliferation.

Fibroblasts are dependent on growth factors only during the pre-commitment G_1 phase of the cell cycle (Zetterberg & Larsson 1985; Pardee 1989): post-commitment $S/G_2/M$ fibroblasts complete their cycle on schedule in the absence of serum and thus do not require any growth factors. We therefore asked whether IGF-I could suppress c-Myc-induced apoptosis in these post-commitment fibroblasts, a time in the cell cycle when the cytokine has no identifiable role in promoting cell proliferation. We observed that activation of c-Myc in serum-deprived post-commitment S/G2 cells induces the immediate onset of apoptosis and this apoptosis is inhibited by either IGF-I or PDGF. Thus, anti-apoptotic cytokines suppress apoptosis in situations where they exert no detectable mitogenic effect.

Finally, we are able to demonstrate that the anti-apoptotic effects of IGF-I are evident even in cells treated with cycloheximide or actinomycin D, which inhibit protein and RNA synthesis respectively. Thus, the anti-apoptotic effect of IGF-I does not require *de novo* gene expression or protein synthesis. This is in clear contrast to the mitogenic programme implemented by IGF-I, which does require *de novo* expression of a panoply of immediate-early growth response genes such as c-*fos*, c-*jun* and *erg*-1. Interestingly, PDGF does not exert any measurable anti-apoptotic effect in the presence of cycloheximide or actinomycin D, indicating that its protective effects do require expression of new gene products. Preliminary data suggest that PDGF may act by inducing IGF-I expression in cells.

In summary, therefore, we can clearly discriminate between the role of IGF-I and PDGF as mitogens and their roles as anti-apoptoptic factors.

5. SIGNIFICANCE OF THE 'DUAL SIGNAL' MODEL FOR CONTROL OF CELL GROWTH

We have demonstrated that c-Myc activates a transcriptional programme within cells, one outcome of which is to establish the capacity for apoptotic cell suicide. This programme is always present in cells that express c-Myc and, because c-Myc appears essential for cell proliferation, is thus always present in proliferating cells. In order to survive entry into cycle, this apoptotic programme must be continuously forestalled by the action of anti-apoptotic cytokines (or, perhaps, anti-apoptotic activities such as that provided by the oncogene *bcl*-2).

The obligatory coupling through c-Myc of the two contradictory programmes of cell proliferation and cell death appears paradoxical. However, the dependence of proliferating cells upon two independent signals, one for proliferation and one for survival, provides a powerful innate mechanism to suppress carcinogenesis (Evan *et al.* 1992; Evan & Littlewood 1993; Harrington *et al.* 1994*b*) which, as discussed above, is a major risk sustained by physically large, long-lived multicellular organisms such as man. The coupling of mitogenic and apoptotic pathways means

that any growth-promoting lesion (e.g. activated oncogene) will be lethal for the affected cell and its progeny as soon as they outgrow their supply of paracrine survival factors. Suppression of neoplastic transformation thus becomes hardwired into the way metazoan cells proliferate. Only the rapid acquisition of a compensating mutation that suppresses cell death (e.g. deregulation of *bcl*-2 or perhaps autocrine activation of the IGF-I signalling pathway) will enable survival of the affected clone. The chance of such simultaneous double mutations occurring are so small that cancer becomes exceedingly rare. Clearly, the model predicts that anti-apoptotic lesions will be common, if not ubiquitous, in tumour cells. Identification and characterization of such lesions, and the molecular basis for their action, should provide an exciting and effective set of novel targets for future therapeutic intervention in cancer.

REFERENCES

Amati, B., Brooks, M., Levy, N., Littlewood, T., Evan, G. & Land, H. 1993 Oncogenic activity of the c-Myc protein requires dimerisation with Max. *Cell* **72**, 233–245.

Amati, B., Dalton, S., Brooks, M., Littlewood, T., Evan, G. & Land, H. 1992 Transcriptional activation by c-Myc oncoprotein in yeast requires interaction with Max. *Nature, Lond.* **359**, 423–426.

Askew, D., Ashmun, R., Simmons, B. & Cleveland, J. 1991. Constitutive c-*myc* expression in IL-3-dependent myeloid cell line suppresses cycle arrest and accelerates apoptosis. *Oncogene* **6**, 1915–1922.

Bennett, M., Anglin, A., McEwan, J., Jagoe, R., Newby, A. & Evan, G. 1993 Inhibition of vascular smooth muscle cell proliferation *in vitro* and *in vivo* by c-*myc* antisense oligonucleotides. *J. Clin. Invest.* **93**, 820–828.

Bissonnette, R., Echeverri, F., Mahboubi, A. & Green, D. 1992 Apoptotic cell death induced by c-*myc* is inhibited by *bcl*-2. *Nature, Lond.* **359**, 552–554.

Dean, M., Levine, R.A., Ran, W., Kindy, M.S., Sonenshein, G.E. & Campisi, J. 1986 Regulation of c-*myc* transcription and mRNA abundance by serum growth factors and cell contact. *J. biol. Chem.* **261**, 9161–9166.

DePinho, R.A., Schreiber-Agus, N. & Alt, F.W. 1991 *myc* family oncogenes in the development of normal and neoplastic cells. *Adv. Cancer Res.* **57**, 1–46.

Dyall, S.D. & Cory, S. 1988 Transformation of bone marrow cells from E mu-myc transgenic mice by Abelson murine leukemia virus and Harvey murine sarcoma virus. *Oncogene Res.* **2**, 403–409.

Eilers, M., Picard, D., Yamamoto, K.R. & Bishop, M.J. 1989 Chimaeras of Myc oncoprotein and steroid receptors cause hormone-dependent transformation of cells. *Nature, Lond.* **340**, 66–68.

Evan, G. & Littlewood, T. 1993 The role of c-*myc* in cell growth. *Curr. Opin. Genet. Dev.* **3**, 44–49.

Evan, G., Wyllie, A., Gilbert, C., Littlewood, T., Land, H., Brooks, M., Waters, C., Penn, L. & Hancock, D. 1992 Induction of apoptosis in fibroblasts by c-*myc* protein. *Cell* **63**, 119–125.

Fanidi, A., Harrington, E. & Evan, G. 1992 Cooperative interaction between c-*myc* and *bcl*-2 proto-oncogenes. *Nature, Lond.* **359**, 554–556.

Gu, W., Bhatia, K., Magrath, I., Dang, C. & Dalla-Favera, R. 1994 Binding and suppression of the myc transcriptional activation domain by p107. *Science, Wash.* **264**, 251–254.

Harrington, E., Fanidi, A., Bennett, M. & Evan, G. 1994*a* Modulation of Myc-induced apoptosis by specific cytokines. *EMBO J.* (In the press.)

Harrington, E., Fanidi, A. & Evan, G. 1994*b* Oncogenes and cell death. *Curr. Opin. Genet. Dev.* **4**, 120–129.

Heikkila, R., Schwab, G., Wickstrom, E., Loke, S.L., Pluznik, D. H., Watt, R. & Neckers, L.M. 1987 A c-myc antisense oligodeoxynucleotide inhibits entry into S phase but not progress from G0 to G1. *Nature, Lond.* **328**, 445–449.

Kato, G.J., Barrett, J., Villa, G.M. & Dang, C.V. 1990 An amino-terminal c-myc domain required for neoplastic transformation activates transcription. *Molec. Cell. Biol.* **10**, 5914–5920.

Kretzner, L., Blackwood, E. & Eisenman, R. 1992 Myc and Max possess distinct transcriptional activities. *Nature, Lond.* **359**, 426–429.

Langdon, W.Y., Harris, A.W. & Cory, S. 1988 Growth of E mu-myc transgenic B-lymphoid cells in vitro and their evolution toward autonomy. *Oncogene Res.* **3**, 271–279.

Loke, S.L., Stein, C., Zhang, X., Avigan, M., Cohen, J. & Neckers, L.M. 1988 Delivery of c-*myc* antisense phosphorothioate oligodeoxynucleotides to hematopoietic cells in culture by liposome fusion: specific reduction in c-myc protein expression correlates with inhibition of cell growth and DNA synthesis. *Curr. Top. Microbiol. Immunol.* **141**, 282–289.

Neiman, P.E., Thomas, S.J. & Loring, G. 1991 Induction of apoptosis during normal and neoplastic B-cell development in the bursa of Fabricius. *Proc. natn. Acad. Sci. U.S.A.* **88**, 5857–61.

Pardee, A.B. 1989 G1 events and regulation of cell proliferation. *Science, Wash.* **246**, 603–608.

Prochownik, E.V., Kukowska, J. & Rodgers, C. 1988 c-*myc* Antisense Transcripts Accelerate Differentiation and Inhibit G1 Progression in Murine Erythroleukemia Cells. *Molec. Cell. Biol.* **8**, 3683–3695.

Spencer, C.A. & Groudine, M. 1991 Control of c-myc regulation in normal and neoplastic cells. *Adv. Cancer Res.* **56**, 1–48.

Stone, J., de Lange, T., Ramsay, G., Jakobvits, E., Bishop, J.M., Varmus, H. & Lee, W. 1987 Definition of regions in human c-*myc* that are involved in transformation and nuclear localization. *Molec. Cell. Biol.* **7**, 1697–1709.

Vaux, D.L., Cory, S. & Adams, J.M. 1988 *Bcl*-2 gene promotes haemopoietic cell survival and cooperates with c-*myc* to immortalize pre-B cells. *Nature, Lond.* **335**, 440–442.

Wagner, A.J., Small, M.B. & Hay, N. 1993 Myc-mediated apoptosis is blocked by ectopic expression of bcl-2. *Molec. Cell. Biol.* **13**, 2432–2440.

Waters, C., Littlewood, T., Hancock, D., Moore, J. & Evan, G. 1991 c-*myc* protein expression in untransformed fibroblasts. *Oncogene* **6**, 101–109.

Wickstrom, E.L., Bacon, T.A., Gonzalez, A., Lyman, G.H. & Wickstrom, E. 1989 Anti-c-myc DNA increases differentiation and decreases colony formation by HL-60 cells. *In Vitro Cell Devl Biol.* **25**, 297–302.

Zetterberg, A. & Larsson, O. 1985 Kinetic analysis of regulatory events in G1 leading to proliferation of quiescence of Swiss 3T3 cells. *Proc. natn. Acad. Sci. U.S.A.* **82**, 5365–5369.

8

The role of the p53 protein in the apoptotic response

D. P. LANE[1], XIN LU[3], TED HUPP[1] AND P. A. HALL[2]

[1]*Cancer Research Campaign Laboratories, University of Dundee, Dundee DD1 4HN, U.K.*
[2]*Department of Pathology, University of Dundee Medical School, Dundee DD1 9SY, U.K.*
[3]*Ludwig Institute for Cancer Research, St Mary's Hospital Medical School, London W2 3PG, U.K.*

SUMMARY

When mammalian cells or tissues are exposed to DNA damaging agents a programmed cell death pathway is induced as well as a cell cycle arrest. In mice in which the p53 gene has been inactivated by homologous recombination this response is profoundly diminished. These mice develop normally so that developmentally induced apoptotic events do not require p53. The p53 gene product is a 393 amino acid nuclear protein that binds specifically to DNA and can act as a positive transcription factor. High levels of p53 can induce the transcription of gene products involved in the cell cycle arrest and apoptotic pathway. The p53 proteins activity is very tightly controlled both by allosteric regulation of its DNA binding function and by regulation of the protein's stability. These results are discussed in the context of the mutations in p53 found in human tumours and their implications for the treatment of the disease by the use of radiation and chemotherapeutic agents that target DNA.

1. INTRODUCTION

The normal development of multicellular eukaryotic organisms requires an ordered process of cell divisions and, as has been appreciated more recently, programmed cell deaths. These cell death processes are tightly controlled both spatially and temporally. Elegant work in the nematode model has emphasized the precision of this process and identified critical gene products required for its execution. It has become evident from careful observations on cells in culture and from the regulation of cell number in adult tissues that programmed cell death is a normal mechanism of tissue homeostasis. Programmed cell death is often seen to occur at very high levels in tumours and is rapidly induced in response to hostile environmental or internal signals, such as the removal of polypeptide growth factors or the exposure of cells to DNA damaging agents. As the gene products required for the induction of apoptosis and for its regulation have been identified in mammalian cells it has been striking how many have been previously identified either as tumour suppressor genes or as oncogenes. This suggests that breakdown of the control of apoptosis may be a key step in the development of cancer. In support of this one of the genes most often found to be mutant in human tumours, the p53 tumour suppressor gene, has been found to be essential for the apoptotic response to DNA damage.

2. RESULTS AND DISCUSSION

(a) p53 and apoptosis

In many human tumours the p53 gene has been inactivated. Typically both alleles are affected, one allele is normally lost by a gross chromosomal deletion while the other allele typically has sustained a point mutation (Hollstein *et al.* 1991). These point mutant proteins continue to be expressed by the tumour cell and indeed may accumulate to high levels (Iggo *et al.* 1990). The point mutations in p53 are commonly found in the central domain of the protein that has recently been shown to constitute the sequence specific DNA binding domain of the molecule (Pavletich *et al.* 1993). In 1990 Moshe Oren and his colleagues discovered that one of these point mutant p53 proteins was in fact a temperature-sensitive mutant (Michalovitz *et al.* 1990). This mutation allowed rapid progress in understanding the function of the protein. When this gene is introduced with an activated ras gene to primary rat fibroblasts at 37°C the cells that express both the ts mutant p53 and the ras gene become immortal and transformed. These cells can be cultured continuously at 37°C, however when they are shifted down to 32°C the cells cease growing and instead arrest, principally at the G1-S phase border of the cell cycle. This arrest coincides with the activity of the ts mutant p53 protein. It is inactive as a transcription factor at 37°C but instead acts as a transforming gene probably by a dominant negative mechanism, whereby it complexes to, and inactivates, the transformed cells own p53 protein (Shaulian *et al.* 1992). As a result of a conformational shift the ts mutant protein regains wild-type activity at 32°C and stimulates the transcription of genes whose products ultimately result in growth arrest. Although an arrest response is seen in ras-gene-transformed fibroblasts this is not the case in other cell types. In some of these reactivation of the mutant p53 by temperature shift results in a dramatic apoptotic response. This programmed cell death response to the ts mutant p53 has been seen in a wide variety of different cell types. It was first shown by Oren and his colleagues working with the M1 mouse myeloid leukaemia cell line (Yonish-Rouach *et al.* 1991) and subsequently seen by other groups working with adenovirus-E1A-transformed cells (Debbas & White 1993; White 1993). The response shows all the characteristic features of apoptosis at both the morphological and the biochemical level. Importantly the apoptotic response of the M1 cells could be blocked by the IL-6 polypeptide growth survival factor and the response in the E1A-transformed cells can be opposed by the action of the known anti-apoptotic genes bcl-2 and adenovirus E1b 19K.

These studies clearly demonstrated that the p53 gene product could induce apoptosis but they did not reveal how this signal might normally be triggered, nor establish the physiological role of the process. The recent production of mice in which both copies of the p53 gene have been inactivated by homologous recombination has allowed a critical test of the role of this gene product in apoptosis. The results have been dramatic as the p53 null mice develop normally to adulthood but are extraordinarily susceptible to the development of cancer (Donehower *et al.* 1992; Harvey *et al.* 1993). When thymocytes (Clarke *et al.* 1993; Lowe *et al.* 1993) or intestinal epithelial cells (Merritt *et al.* 1994) from these animals are exposed *in vivo* to ionising radiation they are found to be extremely resistant to DNA damage induced apoptosis in contrast to the cells of control littermates who possess one or two copies of an intact p53 gene. This phenotype indicates that the p53 protein is a key determinant of the apoptotic response to DNA damage. It is reasonable to assume that it is the loss of this response to damage, induced spontaneously by external environmental agents or internal processes, that accounts for the tumour susceptible phenotype of these animals. This model would also go some way towards explaining the high frequency of p53 mutations in human tumours because inactivation of p53 function would permit the growth of cells with abberant DNA. It further suggests that tumours in which p53 function is ablated may be more resistant to some types of cancer therapy (Lane 1992, 1993).

(b) *The induction of p53 by DNA damage*

The p53 protein normally has a very short half life and although p53 mRNA is present in all adult tissues examined the level of the protein product is so low as to be virtually undetectable. In 1984 it was found that p53 protein levels rose dramatically in cultured fibroblasts exposed to UV light (Maltzman & Czyzyk 1984). The mechanism responsible for this increase in p53 protein appeared to post-translational stabilization since no increase in mRNA was apparent but the p53 protein produced by the damaged cells had a much longer half life. The significance of these observations was not fully appreciated at the time but they now have been confirmed and extended by many groups (Kastan *et al.* 1991; Kuerbitz *et al.* 1992;

Lu *et al.* 1992; Fritsche *et al.* 1993; Lu & Lane 1993). The response is clearly physiological as when human skin is exposed to quite low doses of UV light (recreational mild sunburn) a dramatic increase in p53 protein is seen in the cells of the epidermis and dermis (Hall *et al.* 1993). The increase in p53 protein detected by immunohistochemistry is confined to the cell nucleus and is associated with increased staining for PCNA, the proliferating cell nuclear antigen that is required both for replicative DNA synthesis and DNA repair. In this tissue no proliferative response was seen to radiation; rather it seems as if there may be a cell cycle arrest and repair response. Careful examination of the kinetics of the p53 response to different DNA damaging agents and the study of the effects of drugs that influence repair rates suggests a very close coupling between the actual lesions in the DNA and the increase in p53 protein stability (Lu & Lane 1993). In support of this we recently showed that the accumulation of p53 would occur in cells which had been exposed to a restriction enzyme in the presence of the porating agent streptolysin O (Lu & Lane 1993). Therefore pure double strand breaks in DNA are able to lead to the accumulation of p53. The mechanism underlying the regulation of p53 degradation by exposure of cells to DNA damage is still unresolved. This is a key point for further study as the accumulation of mutant p53 in tumour cells may have the same fundamental basis as the accumulation of the wild-type protein in normal cells exposed to DNA damaging agents.

(c) *The p53 that accumulates in cells exposed to DNA damage is transcriptionally active*

The p53 protein binds sequence specifically to DNA. The protein appears to bind as a tetramer and a consensus recognition sequence has been defined. The DNA binding domain in the central region of the p53 protein is flanked on the N-terminal side by a acidic domain that can act as a transcriptional transactivator element while the C terminus of the protein contains the regions required for oligomerization and nuclear transport. If a p53 consensus binding sequence is placed upstream of a minimal promoter and a reporter gene then transcription of the reporter gene will be p53 dependent. This was found to be the case using a line of prostate-derived cells into which a p53 responsive CAT gene had been introduced (Lu & Lane 1993). These cells contain very low levels of wild-type p53 protein transcribed from their endogenous normal p53 gene. However, when such cells are irradiated by UV light a dramatic increase in p53-dependent transcription of the reporter gene is seen. A number of cellular genes have recently been identified whose transcription is p53 dependent. These include the gene WAF-1 (El-Deiry *et al.* 1993) whose protein product (Harper *et al.* 1993) is a potent inhibitor of the Cyclin/cdk-2 protein kinase. This suggests a very attractive route by which at least some of the biological effects of p53 could be exerted. Inhibition of the cyclin/cdk2 complex would arrest the cell cycle and in some cells this arrest may provoke an apoptotic response. One mechanism by which this pathway could function is that inactivation of cyclin/cdk2 complexes by WAF-1 would result in accumulation of the dephosphorylated form of the retinoblastoma protein, p105 Rb. This form of Rb blocks the cell cycle in G-1 because it complexes to and inactivates the transcription factor E2F (Chellappan *et al.* 1991). E2F transcriptional activity is required for the syntheses of many S-phase proteins such as the replicative DNA polymerase. This model is attractive as we recently found that the cell cycle block imposed by the temperature sensitive mutant p53 can be by-passed by introduction of the adenoviral E1A and Papilloma virus E7 genes (Vousden *et al.* 1993). The products of these genes constitutivly activate E2F because they bind to and inactivate the dephosphorylated form of p105 Rb.

(d) p53 is regulated by an allosteric control mechanism

Because the activation of wild-type p53 as a transcription factor can result in the induction of cell death it is expected that p53 function will be tightly controlled. One mechanism of importance is clearly the regulation of the proteins stability discussed above. Work on the biochemical activity of p53 as a sequence specific DNA binding protein has revealed another important form of regulation (Hupp *et al.* 1992, 1993). When p53 is produced in bacterial or insect cell expression systems most of the protein is biochemically inactive in DNA binding assays. This latent form of the protein can be activated by an allosteric shift that is induced by phosphorylation of the penultimate amino acid of the protein by casein kinase 2. The latent form is maintained by the activity of the last 30 amino acids of the protein, the so called negative regulatory domain. Simply removing this region of the protein will result in constitutive activation of p53 as a DNA binding protein. The allosteric activation of p53 can also be brought about by non physiological events such as the binding of the monoclonal antibody PAb421 to the C terminus or the action of the bacterial chaperone protein dnaK (Hupp *et al.* 1992, 1993). Of great interest has been the discovery that certain point mutant p53 proteins can be activated to DNA binding forms only by dnaK and PAb421 but not by phosphorylation. Such mutant proteins may represent a powerful target for the development of novel therapeutics that would rescue mutant p53 proteins to induce apoptosis in tumour cells (Hupp *et al.* 1993).

(e) Viruses and apoptosis

The p53 protein was first discovered because it formed a tight protein complex with the viral oncogene product of the SV40 virus, the SV40 large T antigen (Lane & Crawford 1979). Many other DNA viruses of diverse evolutionary origin also produce oncogene products that interact with p53 suggesting that this is very important for this group of viruses. When the viral proteins bind to p53 they inactivate its function as a transcription factor (Mietz *et al.* 1992). This suggests that the viruses have evolved to inactivate p53 function. This can be understood if the viral life cycle would normally induce a p53-dependant apoptotic response. By blocking p53 function the infected cell will survive longer allowing the virus a longer period for propagation and possibly reducing the vulnerability of the infected cell to external inducing agents. The signal that the virus creates that would normally trigger the p53 pathway is not yet clear though an attractive possibility is that the signal is generated by the replication of the viral genome within the infected cell nucleus.

3. CONCLUSIONS

The p53 protein function in mediating the apoptotic response to DNA damage acts to protect the organism from the development of cancer. This response is highly regulated and its modulation may improve the therapeutic affect of many anti cancer agents. Tumours in which p53 function has been lost by point mutations in the p53 gene may still be susceptible to therapy if a practical route to rescuing some wild-type function from these mutant proteins can be found.

REFERENCES

Chellappan, S.P., Hiebert, S., Mudryj, M., Horowitz, J.M. & Nevins, J.R. 1991 The E2F transcription factor is a cellular target for the RB protein. *Cell* **65**, 1053–1061.

Clarke, A.R., Purdie, C.A., Harrison, D.J., Morris, R.G., Bird, C.C., Hooper, M.L. & Wyllie, A.H. 1993 Thymocyte apoptosis induced by p53-dependent and independent pathways. *Nature, Lond.* **362**, 849–852.

Debbas, M. & White, E. 1993 Wild-type p53 mediates apoptosis by E1A, which is inhibited by E1B. *Genes Dev.* **7**, 546–554.

Donehower, L.A., Harvey, M., Slagle, B.L., McArthur, M.J., Montgomery, C.A., Butel, J.S. & Bradley, A. 1992 Mice deficient for p53 are developmentally normal but susceptible to spontaneous tumours. *Nature, Lond.* **356**, 215–221.

El-Deiry, W.S., Tokino, T., Velculescu, V.E., Levy, D.B., Parsons, R., Trent, J.M., Lin, D., Mercer, W.E., Kinzler, K.W. & Vogelstein, B. 1993 WAF1, a potential mediator of p53 tumor suppression. *Cell* **75**, 817–825.

Fritsche, M., Haessler, C. & Brandner, G. 1993 Induction of nuclear accumulation of the tumor-suppressor protein p53 by DNA-damaging agents. *Oncogene* **8**, 307–318.

Hall, P.A., McKee, P.H., Menage, H.d.P., Dover, R. & Lane, D.P. 1993 High levels of p53 proteins in UV-irradiated normal human skin. *Oncogene* **8**, 203–207.

Harper, J.W., Adami, G.R., Wei, N., Keyomarsi, K. & Elledge, S.J. 1993 The p21 cdk-interacting protein Cip1 is a potent inhibitor of G1 cyclin-dependent kinases. *Cell* **75**, 805–816.

Harvey, M., McArthur, M.J., Montgomery, C.J., Bradley, A. & Donehower, L.A. 1993 Genetic background alters the spectrum of tumors that develop in p53-deficient mice. *FASEB J.* **7**, 938–943.

Hollstein, M., Sidransky, D., Vogelstein, B. & Harris, C. 1991 p53 mutations in human cancer. *Science, Wash.* **253**, 49–53.

Hupp, T.R., Meek, D.M., Midgley, C.A. & Lane, D.P. 1993 Activation of the cryptic DNA binding function of mutant forms of p53. *Nucl. Acids Res.* **21**, 3167–3174.

Hupp, T.R., Meek, D.W., Midgley, C.A. & Lane, D.P. 1992 Regulation of the specific DNA binding function of p53. *Cell* **71**, 875–886.

Iggo, R., Gatter, K., Bartek, J., Lane, D. & Harris, A.L. 1990 Increased expression of mutant forms of p53 oncogene in primary lung cancer. *Lancet* **335**, 675–679.

Kastan, M.B., Onyekwere, P., Sidransky, D., Vogelstein, B. & Craig, R.W. 1991 Participation of p53 protein in the cellular response to DNA damage. *Cancer Res.* **51**, 6304–6311.

Kuerbitz, S.J., Plunkett, B.S., Walsh, W.V. & Kastan, M.B. 1992 Wild-type p53 is a cell cycle checkpoint determinant following irradiation. *Proc. natn. Acad. Sci. U.S.A.* **89**, 7491–7495.

Lane, D.P. 1992 p53, guardian of the genome. *Nature, Lond.* **358**, 15–16.

Lane, D.P. 1993 Cancer. A death in the life of p53. *Nature, Lond.* **362**, 786–787.

Lane, D.P. & Crawford, L.V. 1979 T-antigen is bound to host protein in SV40-transformed cells. *Nature, Lond.* **278**, 261–263.

Lowe, S.W., Schmitt, E.M., Smith, S.W., Osborne, B.A. & Jacks, T. 1993 p53 is required for radiation-induced apoptosis in mouse thymocytes. *Nature, Lond.* **362**, 847–849.

Lu, X. & Lane, D.P. 1993 Differential induction of transcriptionally active p53 following UV or ionising radiation: defects in chromosome instability syndromes. *Cell* **75**, 765–778.

Lu, X., Park, S.H., Thompson, T.C. & Lane, D.P. 1992 ras-induced hyperplasia occurs with mutation of p53, but an activated ras and myc together can induce carcinoma without p53 mutation. *Cell* **70**, 153–161.

Maltzman, W. & Czyzyk, L. 1984 UV irradiation stimulates levels of p53 cellular tumor antigen in nontransformed mouse cells. *Molec. Cell. Biol.* **4**, 1689–1694.

Merritt, A.J., Potten, C.S., Kemp, C.J., Hickman, J.A., Balmain, A., Lane, D.P. & Hall, P.A. 1994 The role of p53 in spontaneous and radiation-induced apoptosis in the gastrointestinal tract of normal and p53 deficient mice. *Cancer Res.* **54**, 614–617.

Michalovitz, D., Halvey, O. & Oren, M. 1990 Conditional inhibition of transformation and of cell proliferation by a temperature-sensitive mutant of p53. *Cell* **62**, 671–680.

Mietz, J.A., Unger, T., Huibregtse, J.M. & Howley, P.M. 1992 The transcriptional transactivation function of wild-type p53 is inhibited by SV40 large T-antigen and by HPV-16 E6 oncoprotein. *EMBO J.* **11**, 5013–5020.

Pavletich, N.P., Chambers, K.A. & Pabo, C. 1993 The DNA binding domain of p53 contains the four conserved regions and major mutation hot spots. *Genes Dev.* **7**, 2556–2564.

Shaulian, E., Zauberman, A., Ginsberg, D. & Oren, M. 1992 Identification of a minimal transforming domain of p53: negative dominance through abrogation of sequence-specific DNA binding. *Molec. Cell. Biol.* **12**, 5581–5592.

Vousden, K.H., Vojtesek, B., Fisher, C. & Lane, D. 1993 HPV-16 E7 or adenovirus E1A can overcome the growth arrest of cells immortalized with a temperature-sensitive p53. *Oncogene* **8**, 1697–1702.

White, E. 1993 Regulation of apoptosis by the transforming genes of the DNA tumor virus adenovirus. *Proc. Soc. exp. Biol. Med.* **204**, 30–39.

Yonish-Rouach, E., Resnitzky, D., Lotem, J., Sachs, L., Kimchi, A. & Oren, M. 1991 Wild-type p53 induces apoptosis of myeloid leukaemic cells that is inhibited by interleukin-6. *Nature, Lond.* **353**, 345–347.

9

Apoptosis regulated by a death factor and its receptor: Fas ligand and Fas

SHIGEKAZU NAGATA

Osaka Bioscience Institute, 6-2-4 Furuedai, Suita, Osaka 565, Japan

SUMMARY

Homeostasis in vertebrates is tightly regulated by not only proliferation and differentiation of cells, but also cell death or apoptosis (Ellis *et al.* 1991; Raff 1992). Many cytokines bind to their respective receptors to regulate proliferation and differentiation of cells. Our recent studies on the Fas ligand and Fas indicate that they work respectively as a death factor and its receptor and suggest that, in some cases, cell death or apoptosis is regulated by cytokines and their receptors. Here, I present the summary of the Fas/Fas ligand system which has been studied in my laboratory over the past 5 years, and I will discuss its physiological roles.

1. FAS, A RECEPTOR FOR A DEATH FACTOR

In 1989, two groups reported mouse monoclonal antibodies having a cytolytic activity on some human cells (Trauth *et al.* 1989; Yonehara *et al.* 1989). The antigens recognized by these antibodies were designated as Fas antigen (Fas) or APO-1. To assess the function of Fas, we isolated human and mouse Fas cDNAs (Itoh *et al.* 1991; Watanabe-Fukunaga *et al.* 1992*b*). Fas consists of 325 (human) or 306 (mouse) amino acids with a signal sequence at the N-terminus and a transmembrane domain in the middle of the molecule. The subsequent purification of human APO-1 and its molecular cloning (Oehm *et al.* 1992) established its identity with Fas. Fas is a member of the tumour necrosis factor (TNF)/nerve growth factor (NGF) receptor family (Itoh *et al.* 1991; Nagata 1993), which include Fas, two TNF receptors (types I and II), the low-affinity NGF receptor, B cell antigen CD40, T cell antigen OX40, CD27 and 4-1BB, and Hodgkin's lymphoma cell surface antigen CD30 (figure 1). The extracellular regions of the family members are rich in cysteine residues, and can be divided into three to six subdomains. The extracellular region is relatively conserved among members (about 24–30% identity), whereas the cytoplasmic region is not, except for some similarity between Fas and the TNF type I receptor (Itoh *et al.* 1991).

The thymus, heart, liver and ovary abundantly express Fas mRNA (Watanabe-Fukunaga *et al.* 1992*b*). Flow cytometry analysis using anti-Fas antibody indicated that most thymocytes except for double negative (CD4⁻ CD8⁻) thymocytes express Fas (Drappa *et al.* 1993; Ogasawara *et al.* 1993). Activated human T cells and B cells express Fas (Trauth *et al.* 1989), and lymphoblastoid cells transformed with human T cell leukemia virus (HTLV)-1, human immunodeficiency virus (HIV) or Epstein-Barr virus (EBV) highly express Fas (Nagata 1994). Some other tumour cell lines such as human myeloid leukemia, human squamous carcinoma and mouse macrophage cell lines also express Fas, although the expression level is low compared with that of the lymphoblastoid cells. Expression of Fas is up-regulated by interferon γ (IFN-γ) in mouse macrophage BAM3 and fibroblasts L929 cells, or human adenocarcinoma HT-29, or by a combination of IFN-γ and TNFα in human tonsillar B cells.

2. MUTATION OF FAS IN *lpr*-MICE

There is a single gene for Fas in human and mouse chromosomes (Adachi *et al.* 1993). The human Fas gene is located on chromosome 10q24.1 (Inazawa *et al.* 1992), whereas mouse Fas gene is in the region of chromosome 19, which is homologous to human 10q24.1 (Watanabe-Fukunaga *et al.* 1992*b*). Mouse Fas chromosomal gene consists of more than 70 kb, and is split by 9 exons (R. Watanabe-Fukunaga and S. Nagata, unpublished results).

Referring the location of Fas gene to the mouse Genomic Database (GBASE). Fas gene was found to be close to the locus called *lpr* (lymphoproliferation) (Watanabe *et al.* 1991). There are two allelic mutations, *lpr* and *lpr*^cg^. The *lpr*-mice hardly express Fas mRNA in the thymus and liver (Watanabe-Fukunaga *et al.* 1992*a*). Accordingly, flow cytometry could not detect Fas on thymocytes from *lpr*-mice (Drappa *et al.* 1993; Ogasawara *et al.* 1993). A comparison of the Fas gene from *lpr*-mice with that of wild-type mice indicated an insertion of an early transposable element (ETn) in intron 2 of Fas gene (Adachi *et al.* 1993). ETn is a mouse endogenous retrovirus, of which about 1000 copies can be found in

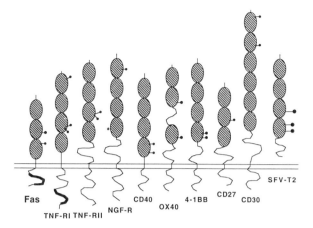

Figure 1. TNF/NGF receptor family. Members of the TNF/NGF receptor family are schematically shown. These include Fas, TNF type I and II receptors, low-affinity NGF receptor, CD40, OX40, 4-1BB, CD27, CD30, and the soluble protein coded by Shope fibroma virus. The slashed regions represent cysteine-rich subdomains. A domain of about 70 amino acids in the cytoplasmic regions of Fas and the type I TNF receptor has some similarity, and it is shown as a bold line. ──● indicates N-glycosylation sites.

the mouse genome (Brulet *et al.* 1983). The ETn carries long terminal repeat (LTR) sequences at both 5′ and 3′ termini, which contains a poly(A) adenylation signal (AATAAA). Inserting the ETn into intron of a mammalian expression vector dramatically, but not completely, reduced the expression efficiency (Adachi *et al.* 1993). These results indicate that in *lpr*-mice, an insertion of an ETn into intron of Fas gene greatly reduces its expression, but this mutation is leaky. In contrast to the *lpr*-mice, *lpr^{cg}* mice express Fas mRNA of normal size as abundantly as the wild-type (Watanabe-Fukunaga *et al.* 1992*a*). However, this mRNA carries a point mutation, which causes a replacement of isoleucine with asparagine in the Fas cytoplasmic region and abolishes the ability of Fas to transduce the apoptotic signal (Watanabe-Fukunaga *et al.* 1992*a*).

3. FAS-MEDIATED APOPTOSIS

(a) *Apoptosis* in vitro

To assess the function of Fas, mouse cell transformants expressing human Fas were established using various mouse cell lines as host (Itoh *et al.* 1991). When the mouse cells expressing human Fas were treated with anti-human Fas antibody, they died within 5 h. Examination of the dying cells under electron microscope revealed extensive condensation and fragmentation of the nuclei, which is characteristic of apoptosis. Chromosomal DNA degraded in a laddered fashion after a 2 h incubation with the anti-Fas antibody. The human Fas expression plasmid has also been introduced into a mouse IL-3 (interleukin-3)-dependent myeloid leukemia FDC-P1 cell line (Itoh *et al.* 1993). Although the transformed cells died due to IL-3 depletion, they did so over 36 h, as observed with the parental cells. On the other

hand, exposure to the anti-human Fas antibody killed the cells within 5 h in the presence of IL-3. From these results, we concluded that Fas actively transduces the apoptotic signal, and the cytolytic anti-human Fas antibody works as agonist.

(b) *Apoptosis* in vivo

We established hamster monoclonal antibodies against mouse Fas, which have cytolytic activity (Ogasawara *et al.* 1993). When this antibody is injected intraperitoneally into mice, the wild-type, but neither *lpr* nor *lpr^{cg}*, mice died within 5–6 h. These results indicate that the lethal effect of the anti-Fas antibody is due to binding of the antibody to the functional Fas in the tissues. The fact that *lpr^{cg}* mice expressing the non-functional Fas are resistant to the lethal effect of the antibody indicates little involvement of the complement system in this killing process. Biochemical analysis of sera showed a specific and dramatic increase of GOT (glutamic oxaloacetic transaminase) and GPT (glutamic pyruvic transaminase) level shortly after injection of the antibody, suggesting the liver injury. Histological analysis indicated focal hemorrhage and necrosis in the liver, whereas dying hepatocytes showed a morphology characteristic of apoptosis under electron microscope (figure 2). These results indicate that individual hepatocytes died by apoptosis. However, as it occurred so rapidly and so widely, granulocytes and macrophages could not phagocytose the apoptotic cells, and the tissues went to the secondary necrosis.

(c) *Apoptotic signal mediated by Fas*

The apoptotic signal through Fas is induced by binding of anti-Fas or anti-APO1 antibody, or the Fas ligand to Fas. The anti-human Fas antibody is an IgM antibody, whereas the anti-APO1 antibody is an IgG$_3$ antibody which tends to aggregate. The F(ab′)$_2$ fragment or other isotypes of the anti-APO1 antibody hardly induces apoptosis (Dhein *et al.* 1992). On the other hand, the cytotoxic activity of the inactive anti-APO1 antibody can be reconstituted by cross-linking the antigen with a secondary antibody or with protein A. These results indicate that oligomerization of at least three Fas molecules is a biologically relevant complex in generating an intracellular signal. As described below, the fact that Fas ligand is a TNF-related molecule which exists as a trimer (Smith & Baglioni 1987), agrees with this hypothesis.

Activation of Fas induces degradation of chromosomal DNA within 3 h, which eventually kills the cells, suggesting that a strong death signal is transduced from Fas. In some cells, activation of Fas alone is not sufficient to induce apoptotic signal. The presence of metabolic inhibitors such as cycloheximide or actinomycin D is required to induce the Fas-dependent apoptosis in these cells (Itoh *et al.* 1991). On the other hand, the activated or transformed T cells can be killed by anti-Fas antibody alone. These results indicate that the signal-transducing machinery for Fas-induced apoptosis is present in most cells, and some cells express

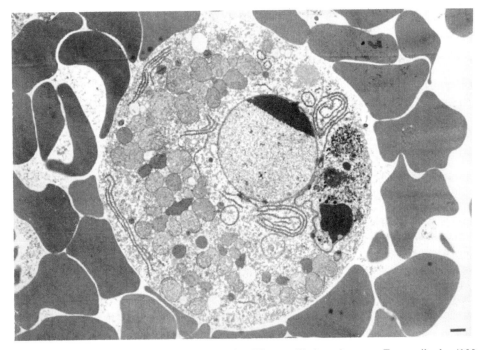

Figure 2. Fas-mediated apoptosis of hepatocytes *in vivo*. The purified anti-mouse Fas antibody (100 μg) was subcutaneously injected into mice and a liver section was examined under a transmission electron microscope. The affected hepatocytes show the condensed and fragmented nuclei characteristic of apoptosis.

a protein(s) which works inhibitory for Fas-mediated apoptosis. In fact, overexpression of oncogene Bcl-2 product which inhibits apoptosis in various system (Korsmeyer 1992) partly inhibited Fas-mediated apoptosis (Itoh *et al*. 1993).

The cytoplasmic domain of Fas consists of 145 amino acids, in which no motif for enzymic activity such as kinases or phosphatase can be found (Itoh *et al*. 1991). However, about 70 amino acids in this region has significant similarity with a part of the cytoplasmic region of the type I TNF receptor (Itoh *et al*. 1991). TNF has numerous biological functions, including cytotoxic and proliferative activities (Old 1985). Tartaglia *et al*. (1991) have shown that the type I TNF receptor is mainly responsible for the cytotoxic activity of TNF, whereas the type II receptor mediates the proliferation signal. Analyses of various mutants in Fas and the type I TNF receptor indicated that the domain conserved between Fas and the type I TNF receptor is essential for apoptotic signal transduction (Itoh & Nagata 1993; Tartaglia *et al*. 1993).

In addition to the signal-transducing domain, Fas carries an inhibitory domain for apoptosis in the C-terminus. That is, a Fas mutant lacking 15 amino acids from the C-terminus was an up-mutant, in which about ten times less anti-Fas antibody than that required for the wild-type Fas was sufficient to induce apoptosis (Itoh & Nagata 1993). Moreover, in L929 cells, activation of Fas alone (without metabolic inhibitors) was sufficient to induce apoptosis. It is possible that association of inhibitory molecule(s) mentioned above or modification of Fas at this region down-regulates the activity of Fas to transduce the apoptotic signal.

4. FAS LIGAND, A DEATH FACTOR

(a) *Identification and purification of Fas ligand*

Rouvier *et al*. (1993) have established a CTL hybridoma cell line (d10S) which has cytotoxic activity against thymocytes from wild-type, but not from *lpr*-mice, suggesting the presence of a Fas ligand on its surface. We prepared a soluble form of Fas (Fas-Fc) by fusing the extracellular region of Fas to the Fc region of human IgG. The fusion protein inhibited the Fas-dependent CTL activity of d10S cells in a dose-dependent manner, and the Fas ligand was detected by FACS on the cell surface of d10S cells using labelled Fas-Fc (Suda & Nagata 1994). A subline of d10S which abundantly expresses the Fas ligand was established by repeated sorting on FACS. After sorting sixteen times, the subline (d10S16) expressed about 100 times more Fas ligand and showed about 100 times more cytotoxic activity than the original d10S cells. The Fas ligand was then purified from d10S16 to homogeneity by affinity chromatographies using Fas-Fc and Con A. The purified Fas ligand had M_r of 40 kDa, and had cytolytic activity specifically against cells expressing Fas (Suda & Nagata 1994), suggesting that a single protein (Fas ligand) is sufficient to induce apoptosis by binding to Fas.

(b) *Molecular properties of the Fas ligand*

A cDNA library was constructed from the sorted subline of d10S cells, and Fas ligand cDNA was isolated by the panning procedure using mFas-Fc (Suda *et al*. 1993). The recombinant Fas ligand expressed in COS cells could kill the cells expressing

Fas by apoptosis. Its amino acid sequence indicated Fas ligand is a type II membrane protein belonging to the TNF family. As shown in figure 3, members of the TNF family include Fas ligand, TNF, lymphotoxin (LT), and ligands for CD40, CD30, CD27 and 4-1BB. All members of this family are type II membrane proteins except for LTα (or TNFβ) which is produced as a soluble cytokine. When Fas ligand was over-produced in COS cells or d10S16 subline, the soluble form of Fas ligand can be found in supernatant (Suda *et al.* 1993). These results suggest that under abnormal conditions, the soluble form of Fas ligand can be produced in the body as found in the TNF system (Old 1985). The tertiary structure of TNF has been extensively studied. It forms an elongated, antiparallel β-pleated sheet sandwich with a jelly-roll topology (Eck & Sprang 1989; Banner *et al.* 1993; Eck *et al.* 1992). The significant conservation of the amino acid sequence among members suggests that others of the family including Fas ligand, have a structure similar to TNF. However, despite the high similarity of the Fas ligand with TNF (about 30% identity of the amino acid sequence level), Fas ligand does not bind to the TNF receptor (Suda *et al.* 1993).

Northern hybridization analysis of rat tissues indicated that Fas ligand is expressed abundantly in the testis, moderately in the small intestines and weakly in the lung. Whereas, little expression of Fas ligand mRNA was observed in the thymus, liver, heart and ovary where Fas is abundantly expressed. In accord with the expression of Fas ligand in the CTL cell line of d10S, activation of splenocytes with phorbol myristic acetate (PMA) and ionomycin strongly induced the expression of Fas ligand mRNA (Suda *et al.* 1993). However, the expression level of

Fas ligand mRNA in thymocytes was relatively weak even after activation with PMA and ionomycin.

5. PHYSIOLOGICAL ROLES OF THE FAS SYSTEM

(a) *Involvement of the Fas system in development of T cells*

As described above, the Fas gene is the structural gene for *lpr*. Because the mice that are homozygous at *lpr* develop lymphadenopathy and suffer from auto-immune disease (Cohen & Eisenberg 1991), it is clear that Fas plays an important role in the development of T cells. However, it remains controversial at which step of T cell development Fas is involved. T cells are killed by apoptosis at least in three steps during their development (Ramsdell & Fowlkers 1990). In the thymus, T cells carrying T cell receptors which do not recognize self-MHC antigens as a restriction element are killed or 'neglected', whereas the T cells recognizing the self antigens are killed by a process called 'negative selection'. Analysis of thymic T cell development in wild-type and *lpr* mice has suggested that the 'neglected' thymocytes escape from apoptosis in the thymus of *lpr* mice, then migrate to the periphery (Zhou *et al.* 1993). On the other hand, Herron *et al.* (1993) and Sidman *et al.* (1992) reported that the development of T cell in the thymus is essentially normal in *lpr* mice. In addition to the thymus, autoreactive mature T cells are deleted in the periphery (Kabelitz *et al.* 1993). Fas is expressed in activated mature T cells (Trauth *et al.* 1989), and the prolonged activation of T cells leads the cells susceptible against cytolytic activity of anti-Fas antibody (Owen-Schaub *et al.* 1992; Klas *et al.* 1993). Because mature T cells from *lpr* mice are resistant against anti-CD3-stimulated suicide, this suggests a role of Fas-mediated apoptosis in the induction of peripheral tolerance and/or in the antigen-stimulated suicide of mature T cells (Bossu *et al.* 1993; Russell & Wang 1993).

(b) *Involvement of the Fas system in CTL-mediated cytotoxicity*

Fas ligand is expressed in some CTL cell lines and in activated splenocytes (Suda *et al.* 1993), suggesting an important role of the Fas system in CTL-mediated cytotoxicity. Two mechanisms for CTL-mediated cyto-toxicity are known (Golstein *et al.* 1991; Podack *et al.* 1991; Apasov *et al.* 1993). The one is a Ca^{2+}-dependent pathway in which perforin and granzymes play an important role. The other pathway is a Ca^{2+}-independent pathway, the mechanism of which is not well understood. It is possible that most Ca^{2+}-independent CTL activity is mediated by the Fas system.

Not only in the thymocytes and lymphocytes, Fas is expressed in other tissues such as the liver, heart and lung (Watanabe *et al.* 1991). Although these organs are rather stable, and no apparent abnormal phenotypes are seen in these tissues of *lpr* mice, Fas may also be

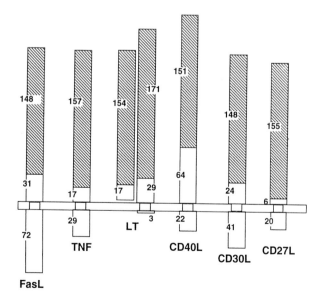

Fig. 3. TNF family. Members of the TNF family are schematically shown. The members include the Fas ligand (FasL), TNF, Lymphotoxin (LT) which consists of LTα and LTβ, CD40 ligand (CD40L), CD30 ligand (CD30L) and CD27 ligand (CD27L). The slashed regions have significant similarity. Numbers indicate the amino acid number of the conserved, the spacer and intracellular regions.

involved in development and/or turnover in these tissues. Because abnormal activation of Fas (administration of anti-Fas antibody) causes severe tissue damage (Ogasawara *et al.* 1993) as described above, it is possible that the Fas system is involved in various human diseases such as fulminant hepatitis. In this regard, it is notable that a particular CTL cell line induces apoptosis in hepatocytes, which leads to fulminant hepatitis (Ando *et al.* 1993; Chisari 1992).

The mutations in Fas (*lpr*) cause lymphadenopathy and autoimmune disease, whereas Fas ligand was found in CTL which kill the tumour cells. These results imply that the Fas/Fas ligand system involved in the T cell development plays an important role in CTL-mediated cytotoxicity. Moreover, it suggests that the killing process of autoreactive T cells in T cell development and the killing process of tumour cells by CTL may proceed by a similar mechanism. As schematically shown in figure 4*a*, autoreactive T cells recognize the self antigens as a complex with MHC which are expressed in the antigen-presenting cells, and may be activated through the T cell receptor. Activation of T cells induces the expression of Fas and Fas ligand, and kill each other. Because *lpr* mice have defects in B cells by producing autoantibodies, the Fas system may also operate to delete the autoreactive B cells in a similar fashion. In the CTL reaction, the target cells such as tumour cells or the cells transformed with virus express the tumour antigen or virus antigen as a complex with MHC. The interaction of CTL with these cells may activate the CTL through the T cell receptor, and induce the Fas ligand gene. The Fas ligand then binds to Fas on the target cells, causing apoptosis (figure 4*b*).

Mice carrying the *gld* mutation show phenotypes similar to *lpr* (Cohen & Eisenberg 1991). From the bone-marrow transplantation experiments between *lpr* and *gld* mice, Allen *et al.* (1990) suggested that *gld* and *lpr* are mutations of an interacting pair of molecules. As shown above, the *lpr* is a mutation in Fas which is the receptor for Fas ligand. Therefore, it is possible that *gld* mice carry mutations in the Fas ligand gene. In fact, we have recently shown that *gld* mice carry a point mutation in the Fas ligand, which inactivates the ability to bind Fas (Takahashi *et al.* 1994).

6. PERSPECTIVES

We demonstrated that Fas ligand is a death factor, and Fas is its receptor. These results indicate that

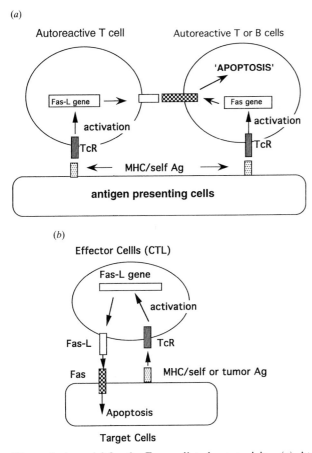

(a)

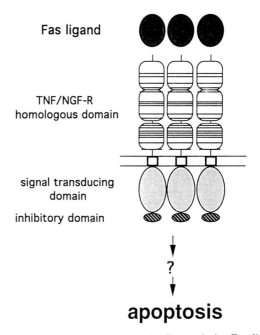

Figure 4. A model for the Fas-mediated cytotoxicity. (*a*) A proposed mechanism for the Fas-mediated peripheral clonal deletion is schematically shown. Antigen-presenting cells express self antigen as complex as MHC, which interacts with T cell receptor in autoreactive T cells and activate them. Activation of T cell induces Fas and Fas ligand gene expression. Interaction of T cells will cause clonal deletion. (*b*) A proposed mechanism for the Fas-mediated cytotoxicity in the CTL system is schematically shown. The target cells express the self, tumour or virus antigen as a complex with MHC, which interacts with the T cell receptor (TcR) on CTL. This interaction activates the CTL, and induces the expression of the Fas ligand (Fas-L) gene. The Fas-L expressed on the cell surface of the CTL then binds to Fas on the target cells, and induces its apoptosis.

Figure 5. Fas-mediated apoptosis. Fas and the Fas ligand are schematically shown. The Fas ligand binds to Fas on the cell surface probably as a trimer, and activates apoptotic signal transduction. In the cytoplasmic region of Fas, a region of about 80 amino acids is responsible for the signal transduction, while the C-terminal domain (about 15 amino acids) inhibits apoptosis.

just as growth factor and its receptor regulate cell proliferation, apoptosis is also regulated by a death factor and its receptor (figure 5). It would be interesting to examine what kinds of signals are transduced through Fas to induce apoptosis. The gain-of-function mutation of the growth factor system causes cellular transformation, whereas the loss-of-function mutation of the Fas system causes lymphadenopathy. In this regard, Fas and the Fas ligand may be considered as tumour suppressor genes. The loss-of-function mutation in the growth factor system causes the disappearance or dysfunction of specific cells. As pointed out above, abnormal activation (gain-of-function) of the Fas or Fas ligand may cause fulminant hepatitis or other diseases such as CTL-mediated autoimmune diseases. If involvement of the Fas system in human diseases is proven, antagonistic antibodies against Fas or Fas ligand, or the soluble form of Fas, could be used in a clinical setting.

I thank Dr O. Hayaishi and Professor C. Weissmann for encouragement and discussion. The work was carried out in a collaboration with Dr T. Suda, Dr J. Ogasawara, Dr T. Takahashi, Dr M Adachi, Dr N. Itoh and Dr R. Watanabe-Fukunaga, and supported in part by Grants-in-Aid from the Ministry of Education, Science and Culture of Japan. I also thank Ms K. Mimura for secretarial assistance.

REFERENCES

Adachi, M., Watanabe-Fukunaga, R. & Nagata, S. 1993 Aberrant transcription caused by the insertion of an early transposable element in an intron of the Fas antigen gene of *lpr* mice. *Proc. natn. Acad. Sci. U.S.A.* **90**, 1756–1760.

Allen, R.D., Marshall, J.D., Roths, J.B. & Sidman, C.L. 1990 Differences defined by bone marrow transplantation suggest that *lpr* and *gld* are mutations of genes encoding an interacting pair of molecules. *J. exp. Med.* **172**, 1367–1375.

Ando, K., Moriyama, T., Guidotti, L.G., Wirth, S., Schreiber, R.D., Schlicht, H.J., Huang, S. & Chisari, F.V. 1993 Mechanisms of class I restricted immunopathology. A transgenic mouse model of fulminant hepatitis. *J. exp. Med.* **178**, 1541–1554.

Apasov, S., Redegeld, F. & Sitkovsky, M. 1993 Cell-mediated cytotoxicity: contact and secreted factors. *Curr. Opin. Immunol.* **5**, 404–410.

Banner, D.W., D'Arcy, A., Janes, W., Gentz, R., Schoenfeld, H.-J., Broger, C., Loetscher, H. & Lesslauer, W. 1993 Crystal structure of the soluble human 55 kd TNF receptor–human TNFβ complex: implication for TNF receptor activation. *Cell* **73**, 431–445.

Bossu, P., Singer, G.G., Andres, P., Ettinger, R., Marshak-Rothstein, A. & Abbas, A.K. 1993 Mature CD4+ T lymphocytes from MRL/lpr mice are resistant to receptor-mediated tolerance and apoptosis. *J. Immunol.* **151**, 7233–7239.

Brulet, P., Kaghad, M., Xu, Y.-S., Croissant, O. & Jacob, F. 1983 Early differential tissue expression of transposon-like repetitive DNA sequences of the mouse. *Proc. natn. Acad. Sci. U.S.A.* **80**, 5641–5645.

Chisari, F.V. 1992 Hepatitis B virus biology and pathogenesis. *Molec. Genet. Med.* **2**, 67–103.

Cohen, P.L. & Eisenberg, R.A. 1991 *Lpr* and *gld*: single gene models of systemic autoimmunity and lymphoproliferative disease. *A. Rev. Immunol.* **9**, 243–269.

Dhein, J., Daniel, P.T., Trauth, B.C., Oehm, A., Möller, P. & Krammer, P.H. 1992 Induction of apoptosis by monoclonal antibody anti-APO-1 class switch variants is dependent on cross-linking of APO-1 cell surface antigens. *J. Immunol.* **149**, 3166–3173.

Drappa, J., Brot, N. & Elkon, K.B. 1993 The Fas protein is expressed at high levels on CD4+CD8+ thymocytes and activated mature lymphocytes in normal mice but not in the lupus-prone strain, MRL *lpr/lpr*. *Proc. natn. Acad. Sci. U.S.A.* **90**, 10340–10344.

Eck, M.J. & Sprang, S.R. 1989 The structure of tumor necrosis factor-α at 2.6 Å resolution. *J. biol. Chem.* **264**, 17595–17605.

Eck, M.J., Ultsch, M., Rinderknecht, E., de Vos, A.M. & Sprang, S.R. 1992 The structure of human lymphotoxin (tumor necrosis factor-β) at 1.9-Å resolution. *J. biol. Chem.* **267**, 2119–2122.

Ellis, R.E., Yuan, J. & Horvitz, H.R. 1991 Mechanisms and functions of cell death. *A. Rev. Cell Biol.* **7**, 663–698.

Golstein, P., Ojcius, D.M. & Young, J.D.-E. 1991 Cell death mechanisms and the immune system. *Immunol. Rev.* **121**, 29–65.

Herron, L.R., Eisenberg, R.A., Roper, E., Kakkanaiah, V.N., Cohen, P.L. & Kotzin, B.L. 1993 Selection of the T cell receptor repertoire in *Lpr* mice. *J. Immunol.* **151**, 3450–3459.

Inazawa, J., Itoh, N., Abe, T. & Nagata, S. 1992 Assignment of the human Fas antigen gene (FAS) to 10q24.1. *Genomics* **14**, 821–822.

Itoh, N. & Nagata, S. 1993 A novel protein domain required for apoptosis: mutational analysis of human Fas antigen. *J. biol. Chem.* **268**, 10932–10937.

Itoh, N., Tsujimoto, Y. & Nagata, S. 1993 Effect of bcl-2 on Fas antigen-mediated cell death. *J. Immunol.* **151**, 621–627.

Itoh, N., Yonehara, S., Ishii, A., Yonehara, M., Mizushima, S., Sameshima, M., Hase, A., Seto, Y. & Nagata, S. 1991 The polypeptide encoded by the cDNA for human cell surface antigen Fas can mediate apoptosis. *Cell* **66**, 233–243.

Kabelitz, D., Pohl, T. & Pechhold, K. 1993 Activation-induced cell death (apoptosis) of mature peripheral T lymphocytes. *Immunol. Today* **338**, 338–340.

Klas, C., Debatin, K.-M., Jonker, R.R. & Krammer, P.H. 1993 Activation interferes with the APO-1 pathway in mature human T cells. *Int. Immunol.* **5**, 625–630.

Korsmeyer, S.J. 1992 Bcl-2 initiates a new category of oncogenes: regulators of cell death. *Blood* **80**, 879–886.

Nagata, S. 1994 Apoptosis-mediating Fas antigen and its natural mutation. In *Apoptosis II, the molecular basis of cell death* (ed. T. D. Tomei & F. C. Cope), pp. 313–326. Cold Spring Harbor Press.

Nagata, S. 1994 Fas and Fas ligand: a death factor and its receptor. *Adv. Immunol.* (In the press).

Oehm, A., Behrmann, I., Falk, W., Pawlita, M., Maier, G., Klas, C., Li-Weber, M., Richards, S., Dhein, J., Trauth, B.C., Ponstingl, H. & Krammer, P.H. 1992 Purification and molecular cloning of the APO-1 cell surface antigen, a member of the tumor necrosis factor/nerve growth factor receptor superfamily: sequence identity with the Fas antigen. *J. biol. Chem.* **267**, 10709–10715.

Ogasawara, J., Watanabe-Fukunaga, R., Adachi, M., Matsuzawa, A., Kasugai, T., Kitamura, Y., Itoh, N., Suda, T. & Nagata, S. 1993 Lethal effect of the anti-Fas antibody in mice. *Nature, Lond.* **364**, 806–809.

Old, L.J. 1985 Tumor necrosis factor (TNF). *Science, Wash.* **230**, 630–632.

Owen-Schaub, L.B., Yonehara, S., Crump III, W.L. & Grimm, E.A. 1992 DNA fragmentation and cell death is selectively triggered in activated human lymphocytes by Fas antigen engagement. *Cell. Immunol.* **140**, 197–205.

Podack, K.R., Hengartner, H. & Lichtenheld, M.G. 1991 A central role of perforin in cytolysis? *A. Rev. Immunol.* **9**, 129–157.

Raff, M.C. 1992 Social controls on cell survival and cell death. *Nature, Lond.* **356**, 397–400.

Ramsdell, F. & Fowlkers, B.J. 1990 Clonal deletion versus clonal anergy: the role of the thymus in inducing self tolerance. *Science, Wash.* **248**, 1342–1348.

Rouvier, E., Luciani, M.-F. & Golstein, P. 1993 Fas involvement in Ca^{2+}-independent T cell-mediated cytotoxicity. *J. exp. Med.* **177**, 195–200.

Russell, J.H. & Wang, R. 1993 Autoimmune *gld* mutation uncouples suicide and cytokine/proliferation pathways in activated, mature T cells. *Eur. J. Immunol.* **23**, 2379–2382.

Sidman, C.L., Marshall, J.D. & Von Boehmer, H. 1992 Transgenic T cell receptor interactions in the lymphoproliferative and autoimmune syndromes of *lpr* and *gld* mutant mice. *Eur. J. Immunol.* **22**, 499–504.

Smith, R.A. & Baglioni, C. 1987 The active form of tumor necrosis factor is a trimer. *J. biol. Chem.* **262**, 6951–6954.

Suda, T. & Nagata, S. 1994 Purification and characterization of the Fas ligand that induces apoptosis. *J. exp. Med.* **179**, 873–878.

Suda, T., Takahashi, T., Golstein, P. & Nagata, S. 1993 Molecular cloning and expression of the Fas ligand: a novel member of the tumor necrosis factor family. *Cell* **75**, 1169–1178.

Takahashi, T., Tanaka, M., Brannan, C.I., Jenkins, N.A., Copeland, N.G., Suda, T. & Nagata, S. 1994 Generalized lymphoproliferative disease in mice, caused by a point mutation in the Fas ligand. *Cell* **76**, 969–976.

Tartaglia, L.A., Ayres, T.M., Wong, G.H.W. & Goeddel, D.V. 1993 A novel domain within the 55 kd TNF receptor signals cell death. *Cell* **74**, 845–853.

Tartaglia, L.A., Weber, R.F., Figari, I.S., Reynolds, C., Palladino Jr., M.A. & Goeddel, D.V. 1991 The two different receptors for tumor necrosis factor mediate distinct cellular responses. *Proc. natn. Acad. Sci. U.S.A.* **88**, 9292–9296.

Trauth, B.C., Klas, C., Peters, A.M.J., Matzuku, S., Möller, P., Falk, W., Debatin, K.-M. & Krammer, P.H. 1989 Monoclonal antibody-mediated tumor regression by induction of apoptosis. *Science, Wash.* **245**, 301–305.

Watanabe, T., Sakai, Y., Miyawaki, S., Shimizu, A., Koiwai, O. & Ohno, K. 1991 A molecular genetic linkage map of mouse chromosome-19, including the *lpr*, *Ly-44*, and *TdT* genes. *Biochem. Genet.* **29**, 325–336.

Watanabe-Fukunaga, R., Brannan, C.I., Copeland, N.G., Jenkins, N.A. & Nagata, S. 1992a Lymphoproliferation disorder in mice explained by defects in Fas antigen that mediates apoptosis. *Nature, Lond.* **356**, 314–317.

Watanabe-Fukunaga, R., Brannan, C.I., Itoh, N., Yonehara, S., Copeland, N.G., Jenkins, N.A. & Nagata, S. 1992b The cDNA structure, expression, and chromosomal assignment of the mouse Fas antigen. *J. Immunol.* **148**, 1274–1279.

Yonehara, S., Ishii, A. & Yonehara, M. 1989 A cell-killing monoclonal antibody (anti-Fas) to a cell surface antigen co-downregulated with the receptor of tumor necrosis factor. *J. exp. Med.* **169**, 1747–1756.

Zhou, T., Bluethmann, H., Eldridge, J., Berry, K. & Mountz, J.D. 1993 Origin of CD4⁻CD8⁻B220⁺ T cells in MRL-*lpr/lpr* mice. *J. Immunol.* **150**, 3651–3667.

10

Insights from transgenic mice regarding the role of *bcl*-2 in normal and neoplastic lymphoid cells

SUZANNE CORY, ALAN W. HARRIS AND ANDREAS STRASSER

The Walter and Eliza Hall Institute of Medical Research, P.O. Royal Melbourne Hospital, Victoria 3050, Australia

SUMMARY

The *bcl*-2 gene was first discovered by molecular analysis of the 14;18 chromosome translocation which is the hallmark of most cases of human follicular lymphoma. To date, it is unique among proto-oncogenes because, rather than promoting cell proliferation, it fosters cell survival. This review summarizes the impact of constitutive *bcl*-2 expression on the development and function of lymphocytes as well as their malignant transformation. Expression of a *bcl*-2 transgene in the B lymphoid compartment profoundly perturbed homeostasis and, depending on the genetic background, predisposed to a severe autoimmune disease resembling human systemic lupus erythematosus. T lymphoid cells from *bcl*-2 transgenic mice were remarkably resistant to diverse cytotoxic agents. Nevertheless, T lymphoid homeostasis was unaffected and tolerance to self was maintained. Expression of high levels of Bcl-2 facilitated the development of B lymphoid tumours but at relatively low frequency and with long latency. Co-expression of *myc* and *bcl*-2, on the other hand, promoted the rapid onset of novel tumours which appeared to derive from a lympho-myeloid stem or progenitor cell. Introduction of the *bcl*-2 transgene into scid mice facilitated the survival and differentiation of pro-B but not pro-T cells, suggesting that a function necessary to supplement or complement the action of Bcl-2 is expressed later in the T than the B lineage. Crosses of the *bcl*-2 transgenic mice with p53$^{-/-}$ mice have addressed whether loss of p53 function and gain of *bcl*-2 function are synergistic for lymphoid cell survival.

1. INTRODUCTION

The 14;18 chromosome translocation typical of human follicular lymphoma (Yunis *et al.* 1987) results from a reciprocal recombination event involving the *bcl*-2 gene and the immunoglobulin heavy chain locus (Tsujimoto *et al.* 1984; Bakhshi *et al.* 1985; Cleary *et al.* 1986). Unlike most other proto-oncogenes, *bcl*-2 does not play a role in cellular proliferation. Instead, it fosters cell survival. This function was first revealed by infection of IL-3-dependent mouse cell lines with a retrovirus engineered to express human Bcl-2 protein (Vaux *et al.* 1988). Such cells normally perish when the growth factor is withdrawn, because they undergo apoptosis (Williams *et al.* 1990), a suicidal process characterized by shrinkage of the cytoplasm, membrane blebbing, chromatin condensation and DNA fragmentation (Wyllie *et al.* 1980). Cells infected with the *bcl*-2 retrovirus did not die after removal of IL-3; although they ceased proliferating, they remained viable for at least two weeks (Vaux *et al.* 1988). Similar effects were subsequently observed for certain cell lines dependent on other growth factors (Nuñez *et al.* 1990).

The *bcl*-2 gene appears to be the mammalian homologue of *ced*-9, which determines cell survival during development of the nematode *Caenorhabditis elegans* (Hengartner *et al.* 1992). Until recently, *bcl*-2

was the only known mammalian cell survival gene. However, an homologous gene has now been identified and shown to encode a protein, Bcl-x$_L$, which also fosters cell survival (Boise *et al.* 1993). The 26 kDa cytoplasmic Bcl-2 protein has a hydrophobic carboxy terminus and is associated with the nuclear envelope, endoplasmic reticulum and mitochondrion (Chen-Levy *et al.* 1989; Monaghan *et al.* 1992; Lithgow *et al.* 1994), where it resides in the outer (Lithgow *et al.* 1994), not the inner membrane as claimed earlier (Hockenbery *et al.* 1990). Bcl-x$_L$ apparently has a similar distribution. Bcl-2 appears to function as a heterodimer with another homologous protein, Bax (Oltvai *et al.* 1993). High levels of Bax inhibit the survival function of Bcl-2. A similar property has been ascribed to Bcl-x$_S$, which is produced by alternative splicing of *bcl*-x transcripts (Boise *et al.* 1993). The biochemical basis for the ability of Bcl-2 and its homologues to regulate cell survival is unknown but its protective function is well conserved, because Bcl-2 preserves mammalian neurons (Garcia *et al.* 1992) and can even spare the cells of *C. elegans* fated to die during ontogeny (Vaux *et al.* 1992; Hengartner & Horvitz 1994).

Mice expressing a *bcl*-2 transgene have been developed by several laboratories to investigate the role of *bcl*-2 in lymphoid development and lymphomagenesis. The transgenes we constructed (Strasser *et al.* 1990*b*) express human *bcl-2* cDNA (Cleary *et al.* 1986) under the control of the intronic enhancer from the immunoglobulin heavy chain locus (Eμ). As expected, most Eμ-*bcl-2* lines expressed the trangene exclusively in the B lymphoid compartment. However, a few expressed it in both B and T cells and one line expressed the transgene in only the T lymphoid compartment. This paper reviews insights derived from analysis of the Eμ-*bcl-2* mice and the progeny of crosses with other mutant mice.

2. BCL-2 AND B LYMPHOID DEVELOPMENT

Expression of the Eμ-*bcl-2* transgene rendered B and T cells remarkably robust, enhancing their longevity *in vitro* in the absence of growth factors and enabling them to survive exposure to diverse cytotoxic agents *in vitro* and *in vivo*, including γ-radiation and cortico-steroids (Sentman *et al.* 1991; Strasser *et al.* 1991*a,b*; Siegel *et al.* 1992). T lymphoid development appeared normal (see below) but B lymphoid homeostasis was profoundly perturbed. The mice displayed a poly-clonal excess of mature, phenotypically normal B cells in all lymphoid organs (McDonnell *et al.* 1989; Strasser *et al.* 1991*b*). Most of these cells were non-cycling but responded to mitogens and growth factors. The imbalance was not confined to B cells, as the number of pre-B cells and Ig-secreting cells was also elevated and serum Ig levels were abnormally high (Strasser *et al.* 1991*b*). Immunization provoked a greatly amplified and prolonged immune response. All of these properties are consistent with an increased lifespan for *bcl-2*-expressing B lymphoid cells.

Our *bcl-2* lines were initially maintained on a mixed genetic background equivalent to (C57BL/6 × SJL)F2. The lines having B lymphoid expression of the transgene proved to be highly prone to develop a fatal auto-immune disease which resembled human systemic lupus erythematosus, being characterized by immune complex glomerulonephritis and high levels of anti-nuclear antibodies (Strasser *et al.* 1991*b*). This autoimmune disease may reflect pathological accumulation of autoreactive antibodies due to the longevity of cells with anti-self reactivity. However, further studies have revealed that the disease has a strong genetic component. Serial crosses of the *bcl-2* transgene on to a C57BL/6 or BALB/c background seem to have eliminated the kidney disease (Strasser *et al.* 1993; A. W. Harrris and M. L. Bath, unpublished results). Thus the onset of the autoimmune disease apparently requires the presence of alleles from the SJL background.

The inference from these results for normal B cell differentiation is that cells with useful antigenic specificities may be triggered to express Bcl-2 to ensure their survival. Conversely, down-regulation of Bcl-2 may allow the demise of cells incapable of interacting with antigen. Consistent with this hypothesis, both B cells undergoing positive selection in germinal centres and circulating B cells express high levels of Bcl-2 protein (Liu *et al.* 1991) and expression of the *bcl-2* transgene partially inhibits deletion of autoreactive B cells in immunoglobulin transgenic mice (Hartley *et al.* 1993).

3. BCL-2 AND T LYMPHOID DEVELOPMENT

Most T lymphocytes which develop in the thymus are doomed to die (Egerton *et al.* 1990), having failed the stringent selection criteria which ensure that only cells bearing useful antigen receptors can mature and emigrate to the periphery. To survive, immature thymocytes must express antigen receptors capable of binding to molecules of the major histocompatibility complex (MHC) on thymic epithelial cells (positive selection). However, those with receptors which bind with high affinity to MHC molecules complexed to self-antigens are censored (negative selection). The self-reactive thymocytes are believed to die by apoptosis, although apoptotic cells are rarely seen within the thymus, probably because they are rapidly phagocytosed. Mature medullary thymocytes contain much more Bcl-2 protein than do immature cortical cells (Pezzella *et al.* 1990; Hockenbery *et al.* 1991), suggesting that modulation of *bcl-2* expression plays an important role in T cell selection.

Surprisingly, however, the *bcl-2* transgene had no major impact on T lymphoid homeostasis. The numbers and relative proportions of all major subpopulations of T cells were normal and thymic involution with age was unaffected (Sentman *et al.* 1991; Strasser *et al.* 1991*a*; Siegel *et al.* 1992). Furthermore, negative selection was apparently still effective, as no self-reactive T cells were detected in the peripheral lymphoid organs, as judged by analysis of the response to endogenous Mls 'superantigens'. Nevertheless, the CD4$^+$8$^+$ cells of *bcl-2* mice were refractory to *in vivo* or *in vitro* treatment with anti-CD3 antibody (Strasser *et al.* 1991*a*; Sentman *et al.* 1991), which kills conventional CD4$^+$8$^+$ cells and is believed to mimic negative selection (Smith *et al.* 1989; Shi *et al.* 1991).

To resolve this paradox, we crossed *bcl-2* transgenic mice with mice in which T cells express a transgene encoding a T-cell receptor recognizing the male HY antigen in the context of the H-2D^b allele of the MHC (von Boehmer 1990). Male H-2D^b anti-HY TCR transgenic mice have a very small thymus, due to deletion of the self-reactive T cells. Expression of *bcl-2* reduced the efficiency of deletion, since *bcl-2*/TCR transgenic male mice accumulated four- to sixfold more thymocytes than TCR transgenic male litter-mates (Strasser *et al.* 1994*b*). Anti-HY TCR-expres-sing cells were also more numerous in the peripheral lymphoid tissues, but these cells expressed abnormally low levels of CD8 co-receptor and were not responsive to the HY antigen. Thus, although *bcl-2* expression hampered deletion of immature self-reactive cells in the thymus, tolerance to self was maintained.

Analysis of female anti-HY mice established that bcl-2 expression prolonged the survival of thymocytes in a non-selecting background. This suggests that upregulation of Bcl-2 could be a consequence of positive selection (Strasser *et al.* 1994*b*).

4. BCL-2 EXPRESSION PROMOTES B- BUT NOT T-LYMPHOID DEVELOPMENT IN SCID MICE

Lymphoid ontogeny depends upon the rearrangement and expression of antigen receptor genes. If productive rearrangement is achieved, a developing B lymphoid cell first expresses a surface μ heavy chain in association with the λ5 and VpreB surrogate light chains (Melchers *et al.* 1993), while an immature T cell initially expresses a TCR β chain complexed to a gp33 polypeptide (Groettrup *et al.* 1993). Signals from these receptors apparently are required both for survival and further differentiation, since lymphoid development is arrested in scid, Rag-1 or Rag-2 deficient mice, which cannot productively rearrange antigen receptor genes (Bosma & Carroll 1991; Mombaerts *et al.* 1992; Shinkai *et al.* 1992).

To test whether *bcl-2* expression could substitute for receptor engagement, we bred scid mice with transgenic mice expressing *bcl-2* in both the T- and B-cell compartments (Strasser *et al.* 1994a). The number of B lymphoid cells in scid/*bcl-2* mice was strikingly higher than in conventional scid littermates. As expected, these cells lacked surface Ig, but they expressed other markers of mature B cells. As most were quiescent, the increase apparently reflected enhanced survival rather than proliferation. These results suggest that upregulation of Bcl-2 may be a signal for positive selection of immature B lymphoid cells in the bone marrow. By analogy with findings for peripheral B cells in germinal centres (Liu *et al.* 1991), we hypothesize that Bcl-2 synthesis is induced as a result of engagement of the antigen receptor. Thus the survival of cells that have achieved productive rearrangement of receptor genes would be ensured.

Surprisingly, T cell development in scid mice was unaffected by the *bcl-2* transgene. Thymocytes were no more numerous in *bcl-2*/scid mice than in scid littermates and died rapidly when cultured (Strasser *et al.* 1994a). There were no detectable peripheral T cells. The failure to remove the block to T cell development was not due to a failure of the pro-T cells to express the transgene, because the level of Bcl-2 detected in sorted (Thy-1$^+$) *bcl-2*/scid thymocytes was comparable to that found in more mature thymocytes from non-scid *bcl-2* mice, which do display enhanced survival (Strasser *et al.* 1991a). Thus, inability to benefit from Bcl-2 (at least at this level of expression) seems to be confined to very early T cells.

Immature T cells acquire responsiveness to Bcl-2 after they display the antigen receptor and, significantly, this ability does not depend on signalling through the TCR. This conclusion was reached by analysis of scid mice which bear the anti-HY TCR transgene in addition to the *bcl-2* transgene (Strasser *et al.* 1994a). To avoid immunological selection by TCR engagement, we analysed female H-2D$^{d/d}$ mice. Scid/ anti-HY TCR mice with this MHC background normally have very few thymocytes. However, the *bcl-2* transgene increased the number of thymocytes 3.5-fold and greatly increased their ability to survive in culture.

Why do immature B and T lymphoid cells differ in their response to Bcl-2? We are struck by the fact that pro-B cells express antigen co-receptor but pro-T cells do not. The B-lymphoid co-receptor, mb-1/B29 (also known as Ig-α/Ig-β (Reth 1992)), is present on pro-B cells in association with surrogate heavy (X) and light chains (VpreB and 5) (Melchers *et al.* 1993). However, the T-lymphoid co-receptor, CD3, is not expressed on pro-T cells; it appears only after rearrangement and expression of TCR genes permits the formation of CD3 complexed with β-gp33 or α-β heterodimers. The T cells in the scid/anti-HY TCR mice bear the transgenic TCR (although it is ineffectual for signalling in the H-2D^d background and cannot promote further differentiation) and would therefore express CD3. Conceivably, therefore, a function essential for realization of the cell survival-promoting activity of Bcl-2 is induced in lymphoid cells via the co-receptor when it first appears at the cell surface (Strasser *et al.* 1994a).

5. BCL-2 AND LEUKAEMIC TRANSFORMATION

To explore the role of *bcl-2* in lymphomagenesis, we determined the frequency of spontaneous tumours over a 12-month period for four independent lines of *bcl-2* mice expressing the transgene in B lymphoid cells and three lines with expression in T lymphoid cells (Strasser *et al.* 1993). The incidence of T lymphomas was barely, if at all, higher than that in non-transgenic mice of the same genetic background. The incidence of B lymphoid tumours was significant (around 10%), although the long latency (more than 40 weeks) argued that somatic mutation had played a major role in their onset. The tumours were predominantly plasmacytomas and novel early B lymphoid tumours. The *myc* gene was commonly rearranged in the plasmacytomas, as reported for *bcl-2* tumours designated large cell lymphoma by others (McDonnell & Korsmeyer 1991). A minority of aggressive human lymphomas display both a *myc* and a *bcl-2* translocation (Mufti *et al.* 1983; Pegoraro *et al.* 1984; de Jong *et al.* 1988; Gauwerky *et al.* 1988; Lee *et al.* 1989).

We concluded from these observations that, by itself, activation of *bcl-2* expression is relatively innocuous. This conclusion is consistent with the indolent course of follicular lymphoma (median survival from diagnosis is more than 10 years) (Horning & Rosenberg 1984) and the fact that a low level of *bcl-2*/J$_H$ recombination can be detected even in non-lymphomatous lymph nodes and tonsils during a strong immune response (Limpens *et al.* 1991). The primary role of *bcl-2* in human follicular lymphoma probably is to enable a cell that has acquired the 14;18 translocation to resist apoptosis in the lymphoid germinal centre. Because the bcl-2 expressing clone can survive adverse circumstances, any chance somatic mutation which conferred a proliferative advantage to a cell within the resistant population would provoke a potent drive toward malignancy (Vaux *et al.* 1988; Strasser *et al.* 1990a).

The presence of *myc* rearrangements within the plasmacytomas arising spontaneously in *bcl-2* transgenic mice and in certain blast crises of follicular lymphoma implied a synergistic role for *myc* and *bcl-2* in the aetiology of these tumours. Direct evidence for synergy was obtained by breeding Eμ-*bcl-2* transgenic mice with Eμ-*myc* mice (Strasser *et al.* 1990*a*). Three-week-old *bcl-2/myc* mice exhibited astoundingly high white blood cell counts (around 5×10^8 cells per ml), and flow cytometry established that the excess cells were predominantly cycling pre-B and B cells. Despite their conspicuous overproduction, these cells were not malignant when injected into non-irradiated histo-compatible mice. Thus, constitutive expression of *bcl-2* and *myc* is insufficient to transform these cell types, as also concluded from our earlier studies on Eμ-*myc* bone marrow cells infected with a *bcl-2* retrovirus (Vaux *et al.* 1988).

The bi-transgenic mice were, however, highly susceptible to tumour development. They all developed transplantable tumours within 7 weeks, significantly faster than their E-*myc* littermates. Surprisingly, these tumours were not the pre-B or B lymphomas typical of Eμ-*myc* mice (Harris *et al.* 1988). Their phenotype suggested that they were derived from a primitive progenitor or stem cell. When cultured *in vitro* with appropriate growth factor combinations, the tumour cells could differentiate down either the macrophage or B lymphoid pathways (Strasser & Cory, unpublished results).

The realization that *myc* can promote apoptosis as well as proliferation has provided insight into the basis for the synergy between *bcl-2* and *myc* in lymphoma development. Myeloid cells and cells constitutively expressing *myc* are highly prone to apoptosis when maintained under sub-optimal growth conditions (Askew *et al.* 1991; Evan *et al.* 1992). Apoptosis of the *myc*-expressing cells can be blocked by antisense *myc* oligonucleotides (Shi *et al.* 1992) or by high levels of intracellular Bcl-2 protein (Fanidi *et al.* 1992; Bissonette *et al.* 1992). Cells mutated to constitutively express *bcl-2* would also presumably have a survival advantage *in vivo* under limiting growth conditions, giving them an enhanced probability of acquiring oncogenic changes that confer a proliferative advantage.

Enhanced susceptibility to apoptosis in the absence of growth factors provides an explanation for the observation that pre-B cells from Eμ-*myc* mice died much faster than normal pre-B cells when cultured *in vitro* in simple medium (Langdon *et al.* 1988). As expected, both the pre-B and B cells from *bcl-2/myc* mice displayed enhanced survival under these conditions but, surprisingly, the *bcl-2/myc* tumour cells did not (Strasser *et al.* 1990*a*). Instead, they died even more rapidly than pre-B cells expressing only the *myc* transgene. The failure of the tumour cells to survive *in vitro* was not due to poor expression of the transgene, as the level of Bcl-2 protein was as high in these cells as in the pre-B and B cells from the *bcl-2/myc* mice (Strasser & Cory, unpublished results). It is apparent from these results that constitutive high expression of Bcl-2 is inadequate for the survival of primitive lympho-myeloid progenitor cells. As the tumour cells are readily transplantable and remain undifferentiated *in vivo*, they must receive an effective survival signal in lymphoid tissues. We surmise that this is due to the presence of a relevant growth factor. We have cultured the tumour cells in an extensive range of growth factors, which includes all the colony stimulating factors, stem cell factor and interleukins 1 to 7 (both singly and in many combinations) but to date have been unable to maintain viable undifferentiated cells. Certain stromal cell lines will, however, sustain them for up to two weeks.

Why is a high concentration of Bcl-2 protein inadequate for the survival of certain cell types under conditions where it so effectively protects closely related cells? We speculate that cell survival may require input from more than one signal transduction pathway, as depicted in figure 1. Signal 1 may elevate the concentration of Bcl-2 (or Bcl-x$_L$), while signal 2 may either elevate the concentration of a function essential for Bcl-2 (or Bcl-x$_L$) activity or reduce the level of inhibitors such as Bax or Bcl-x$_S$. In lymphocytes, we and others have provided data which implies that signal 1 is mediated via activation of the antigen receptor but in other cell types activation of a growth factor receptor may serve this function. The second signal may emanate from a co-receptor, as suggested for pro-B and pro-T cells (Strasser *et al.* 1994*a*); alternatively it may instead involve a second cytokine pathway, as we have speculated for the *bcl-2/myc* progenitor cell tumours.

6. IMPACT OF BCL-2 ON CELL DEATH INDUCED IN LYMPHOCYTES BY DNA DAMAGE

After exposure to DNA damaging agents such as γ-radiation, cells undergo cell cycle arrest (Hartwell & Weinert 1989). Certain mammalian cell types such as fibroblasts then undergo DNA repair but lymphocytes rapidly die by apoptosis (Wyllie 1980). It has recently become clear that the transcription factor encoded by the p53 tumour suppressor gene is a crucial regulator of the cellular response to DNA

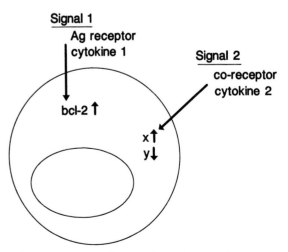

Figure 1. Does cell survival require two signals?

damage. The level of p53 protein rises rapidly after DNA damage (Maltzmann & Czycyk 1984; Kastan *et al.* 1991). Analysis of mice rendered nullizygous for p53 by targeted homologous recombination has established that p53 is essential for both G_1 arrest of fibroblasts (Kastan *et al.* 1992; Livingstone *et al.* 1992) and apoptosis of thymocytes (Lowe *et al.* 1993; Clarke *et al.* 1993). Functional inactivation of p53 is one of the most frequently observed mutations in human tumours. Furthermore, inheritance of either one or two copies of an inactivated p53 gene strongly predisposes toward cancer development (Malkin *et al.* 1990; Donehower *et al.* 1992; Jacks *et al.* 1994). Loss of p53 function is believed to lead to malignant transformation because damaged DNA is neither adequately repaired nor eliminated via apoptosis (Lane 1992).

Because both p53 and *bcl*-2 mutations contribute to neoplastic transformation by blocking apoptosis, we are currently studying the relationship between these two regulators of cell survival and death. By enforcing high levels of Bcl-2 in an erythroleukaemia cell line with inducible wild-type (wt) p53 protein, we have demonstrated that Bcl-2 can block apoptosis induced by wt p53 but fails to prevent the block to proliferation. This observation implies that Bcl-2 acts downstream of the regulator(s) establishing growth arrest or that p53 mediates growth arrest and apoptosis via independent pathways. In a collaboration with Dr Tyler Jacks, we have crossed our *bcl*-2 transgenic mice with p53$^{-/-}$ animals and found that loss of p53 function and gain of *bcl*-2 function are not synergistic for the survival of lymphoid cells. However, *bcl*-2 can inhibit a p53-independent cell death pathway invoked by DNA damage (Strasser *et al.* 1995).

We thank our colleagues in the Molecular Biology Unit, particularly Dr Jerry Adams, for discussions, M. Stanley and M. L. Bath for expert assistance, and J. Parnis and K. Patane for animal husbandry. S.C. is an International Research Scholar of the Howard Hughes Medical Institute. A. S. is supported by fellowships from the Leukemia Society of America and the Swiss National Science Foundation. This work was supported by the National Health and Medical Research Council of Australia, the US National Cancer Institute (CA 43540) and the Howard Hughes Medical Institute (75193-531101).

REFERENCES

Askew, D.S., Ashman, R.A., Simmons, B.C. & Cleveland, J.L. 1991 Constitutive c-*myc* expression in an IL-3-dependent myeloid cell line suppresses cell cycle arrest and accelerates apoptosis. *Oncogene* **6**, 1915–1922.

Bakhshi, A., Jensen, J.P., Goldman, P., Wright, J.J., McBride, O.W., Epstein, A.L. & Korsmeyer, S.J. 1985 Cloning the chromosomal breakpoint of t(14;18) human lymphomas: clustering around J_H on chromosome 14 and near a transcriptional unit on 18. *Cell* **41**, 899–906.

Bissonnette, R.P., Echeverri, F., Mahboubi, A. & Green, D.R. 1992 Apoptotic cell death induced by c-*myc* is inhibited by *bcl*-2. *Nature, Lond.* **359**, 552–554.

Boise, L.H., Gonzalez-Garcia, M., Postema, C.E., Ding, L., Lindsten, T., Turka, L.A., Mao, X., Nuñez, G. & Thompson, C.B. 1993 *bcl-x*, a *bcl*-2-related gene that

functions as a dominant regulator of apoptotic cell death. *Cell* **74**, 597–608.

Bosma, M.J. & Carroll, A.M. 1991 The scid mouse mutant: definition, characterization, and potential uses. *A. Rev. Immunol.* **9**, 323–350.

Chen-Levy, Z., Nourse, J. & Cleary, M.L. 1989 The *bcl*-2 candidate proto-oncogene product is a 24-kilodalton integral-membrane protein highly expressed in lymphoid cell lines and lymphomas carrying the t(14;18) translocation. *Molec. Cell. Biol.* **9**, 701–710.

Clarke, A.R., Purdie, C.A., Harrison, D.J., Morris, R.G., Bird, C.C., Hooper, M.L. & Wyllie, A.H. 1993 Thymocyte apoptosis induced by p53-dependent and independent pathways. *Nature, Lond.* **362**, 849–852.

Cleary, M.L., Smith, S.D. & Sklar, J. 1986 Cloning and structural analysis of cDNAs for *bcl*-2 and a hybrid *bcl*-2/ immunoglobulin transcript resulting from the t(14;18) translocation. *Cell* **47**, 19–28.

de Jong, D., Voetdijk, M.H., Beverstock, G.C., van Ommen, G.J.B., Willemze, R. & Kluin, P.M. 1988 Activation of the c-*myc* oncogene in a precursor-B-cell blast crisis of follicular lymphoma, presenting as composite lymphoma. *New Engl. J. Med.* **318**, 1373–1378.

Donehower, L.A., Harvey, M., Slagle, B.L., McArthur, M.J., Montgomery, C.A.Jr, Butel, J.S. & Bradley, A. 1992 Mice deficient for p53 are developmentally normal but are susceptible to spontaneous tumours. *Nature, Lond.* **356**, 215–221.

Egerton, M., Scollay, R. & Shortman, K. 1990 Kinetics of mature T cell development in the thymus. *Proc. natn. Acad. Sci. U.S.A.* **87**, 2579–2582.

Evan, G.I., Wyllie, A.H., Gilbert, C.S., Littlewood, T.D., Land, H., Brooks, M., Waters, C.M., Penn, L.Z. & Hancock, D.C. 1992 Induction of apoptosis in fibroblasts by c-myc protein. *Cell* **69**, 119–128.

Fanidi, A., Harrington, E.A. & Evan, G.I. 1992 Cooperative interaction between c-*myc* and *bcl*-2 proto-oncogenes. *Nature, Lond.* **359**, 554–556.

Garcia, I., Martinou, I., Tsujimoto, Y. & Martinou, J.-C. 1992 Prevention of programmed cell death of sympathetic neurons by the *bcl*-2 proto-oncogene. *Science, Wash.* **258**, 302–304.

Gauwerky, C.E., Hoxie, J., Nowell, P.C. & Croce, C.M. 1988 Pre-B-cell leukemia with a t(8;14) and a t(14;18) translocation is preceded by follicular lymphoma. *Oncogene* **2**, 431–435.

Groettrup, M., Ungewiss, K., Azogui, O., Palacios, R., Owen, M.J., Hayday, A.C. & von Boehmer, H. 1993 A novel disulfide-linked heterodimer on pre-T cells consists of the T cell receptor β chain and a 33 kd glycoprotein. *Cell* **75**, 283–294.

Harris, A.W., Pinkert, C.A., Crawford, M., Langdon, W.Y., Brinster, R.L. & Adams, J.M. 1988 The Eμ-*myc* transgenic mouse. A model for high-incidence spontaneous lymphoma and leukemia of early B cells. *J. exp. Med.* **167**, 353–371.

Hartley, S.B., Cooke, M.P., Fulcher, D.A., Harris, A.W., Cory, S., Basten, A. & Goodnow, C.C. 1993 Elimination of self-reactive B lymphocytes proceeds in two stages: arrested development and cell death. *Cell* **72**, 1–20.

Hartwell, L.H. & Weinert, T.A. 1989 Checkpoints: controls that ensure the order of cell cycle events. *Science, Wash.* **246**, 629–634.

Hengartner, M.O., Ellis, R.E. & Horvitz, H.R. 1992 *Caenorhabditis elegans* gene *ced-9* protects cells from programmed cell death. *Nature, Lond.* **356**, 494–499.

Hengartner, M.O. & Horvitz, R.H. 1994 C. elegans survival gene *ced*-q encodes a functional homolog of the mammalian proto-oncogene *bcl*-2. *Cell* **76**, 665–676.

Hockenbery, D., Nuñez, G., Milliman, C., Schreiber, R.D. & Korsmeyer, S.J. 1990 *Bcl-2* is an inner mitochondrial membrane protein that blocks programmed cell death. *Nature, Lond.* **348**, 334–336.

Hockenbery, D.M., Zutter, M., Hickey, W., Nahm, M. & Korsmeyer, S. 1991 BCL2 protein is topographically restricted in tissues characterized by apoptotic cell death. *Proc. natn. Acad. Sci. U.S.A.* **88**, 6961–6965.

Horning, S.J. & Rosenberg, S.A. 1984 The natural history of initially untreated low-grade non-Hodgkin's lymphomas. *N. Engl. J. Med.* **311**, 1471–1475.

Jacks, T., Remington, L., Williams, B.O., Schmitt, E.M., Halachmi, S., Bronson, R.T. & Weinberg, R.A. 1994 Tumor spectrum analysis in *p53*-mutant mice. *Curr. Biol.* **4**, 1–7.

Kastan, M.B., Onyekwere, O., Sidransky, D., Vogelstein, B. & Craig, R.W. 1991 Participation of p53 protein in the cellular response to DNA damage. *Cancer Res.* **51**, 6304–6311.

Kastan, M.B., Zhan, Q., El-Deiry, W.S., Carrier, F., Jacks, T., Walsh, W.V., Plunkett, B.S., Vogelstein, B. & Fornace, A.J.Jr 1992 A mammalian cell cycle checkpoint pathway utilizing p53 and GADD45 is defective in ataxia-telangiectasia. *Cell* **71**, 587–597.

Lane, D.P. 1992 p53, guardian of the genome. *Nature, Lond.* **358**, 15–16.

Langdon, W.Y., Harris, A.W. & Cory, S. 1988 Growth of Eμ-*myc* transgenic B-lymphoid cells *in vitro* and their evolution toward autonomy. *Oncogene Res.* **3**, 271–279.

Lee, J.T., Innes, D.J.J. & Williams, M.E. 1989 Sequential *bcl-2* and c-*myc* oncogene rearrangements associated with the clinical transformation of non-Hodgkin's lymphoma. *J. Clin. Invest.* **84**, 1454–1459.

Limpens, J., de Jong, D., van Krieken, J.H.J.M., Price, G.A., Young, B.D., van Ommen, G.-J.B. & Kluin, P.M. 1991 *Bcl-2*/J_H rearrangements in benign lymphoid tissues with follicular hyperplasia. *Oncogene* **6**, 2271–2276.

Lithgow, T., van Driel, R., Bertram, J.F. & Strasser, A. 1994 The protein product of the oncogene *bcl-2* is a component of the nuclear envelope, the endoplasmic reticulum and the outer mitochondrial membrane. *Cell Growth Diff.* **5**, 411–417.

Liu, Y.J., Mason, D.Y., Johnson, G.D., Abbot, S., Gregory, C.D., Hardie, D.L., Gordon, J. & MacLennan, I.C. 1991 Germinal centre cells express bcl-2 protein after activation by signals which prevent their entry into apoptosis. *Eur. J. Immunol.* **21**, 1905–1910.

Livingstone, L.R., White, A., Sprouse, J., Livanos, E., Jacks, T. & Tlsty, T.D. 1992 Altered cell cycle arrest and gene amplification potential accompany loss of wild-type p53. *Cell* **70**, 923–935.

Lowe, S.W., Schmitt, E.M., Smith, S.W., Osborne, B.A. & Jacks, T. 1993 p53 is required for radiation-induced apoptosis in mouse thymocytes. *Nature, Lond.* **362**, 847–849.

Malkin, D., Li, F.P., Strong, L.C., Fraumeni, J.F.J., Nelson, C.E., Kim, D.H., Kassel, J., Gryka, M.A., Bischoff, F.Z., Tainsky, M.A. & Friend, S.H. 1990 Germ line p53 mutations in a familial syndrome of breast cancer, sarcomas, and other neoplasms. *Science, Wash.* **250**, 1233–1238.

Maltzmann, W. & Czyzyk, L. 1984 UV irradiation stimulates levels of p53 cellular tumor antigen in nontransformed mouse cells. *Molec. Cell. Biol.* **4**, 1689–1694.

McDonnell, T.J., Deane, N., Platt, F.M., Nunez, G., Jaeger, U., McKearn, J.P. & Korsmeyer, S.J. 1989 *bcl-2*-immunoglobulin transgenic mice demonstrate extended B cell survival and follicular lymphoproliferation. *Cell* **57**, 79–88.

McDonnell, T.J. & Korsmeyer, S.J. 1991 Progression from lymphoid hyperplasia to high-grade malignant lymphoma in mice transgenic for the t(14;18). *Nature, Lond.* **349**, 254–256.

Melchers, F., Karasuyama, H., Haasner, D., Bauer, S., Kudo, A., Sakaguchi, N., Jameson, B. & Rolink, A. 1993 The surrogate light chain in B-cell development. *Immunol. Today*, **14**, 60–68.

Mombaerts, P., Iacomini, J., Johnson, R.S., Herrup, K., Tonegawa, S. & Papaioannou, V.E. 1992 RAG-1-deficient mice have no mature B and T lymphocytes. *Cell* **68**, 869–877.

Monaghan, P., Robertson, D., Amos, T.A.S., Dyer, M.J.S., Mason, D.Y. & Greaves, M.F. 1992 Ultrastructural localization of BCL-2 protein. *J. Histochem. Cytochem.* **40**, 1819–1825.

Mufti, G.J., Hamblin, T.J., Oscier, D.G. & Johnson, S. 1983 Common ALL with pre-B-cell features showing (8;14) and (14;18) chromosome translocations. *Blood* **62**, 1142–1146.

Nuñez, G., London, L., Hockenbery, D., Alexander, M., McKearn, J.P. & Korsmeyer, S.J. 1990 Deregulated *Bcl-2* gene expression selectively prolongs survival of growth factor-deprived hemopoietic cell lines. *J. Immunol.* **144**, 3602–3610.

Oltvai, Z.N., Milliman, C.L. & Korsmeyer, S.J. 1993 Bcl-2 heterodimerizes *in vivo* with a conserved homolog, Bax, that accelerates programed cell death. *Cell* **74**, 609–619.

Pegoraro, L., Palumbo, A., Erikson, J., Falda, M., Giovanazzo, B., Emanuel, B.S., Rovera, G., Nowell, P.C. & Croce, C.M. 1984 A 14;18 and an 8;14 chromosome translocation in a cell line derived from an acute B-cell leukemia. *Proc. natn. Acad. Sci. U.S.A.* **81**, 7166–7170.

Pezzella, F., Tse, A.G.D., Cordell, J.L., Pulford, K.A.F., Gatter, K.C. & Mason, D.Y. 1990 Expression of the *bcl-2* oncogene protein is not specific for the 14;18 chromosomal translocation. *Am. J. Pathol.* **137**, 225–232.

Reth, M. 1992 Antigen receptors on B lymphocytes. *A. Rev. Immunol.* **10**, 97–121.

Sentman, C.L., Shutter, J.R., Hockenbery, D., Kanagawa, O. & Korsmeyer, S.J. 1991 *bcl-2* inhibits multiple forms of apoptosis but not negative selection in thymocytes. *Cell* **67**, 879–888.

Shi, Y., Bissonnette, R.P., Parfrey, N., Szalay, M., Kubo, R.T. & Green, D.R. 1991 *In vivo* administration of monoclonal antibodies to the CD3 T cell receptor complex induces cell death (apoptosis) in immature thymocytes. *J. Immunol.* **146**, 3340–3346.

Shi, Y., Glynn, J.M., Guilbert, L.J., Cotter, T.G., Bissonnette, R.P. & Green, D.R. 1992 Role for c-myc in activation-induced apoptotic cell death in T cell hybridomas. *Science, Wash.* **257**, 212–214.

Shinkai, Y., Rathbun, G., Lam, K.-P., Oltz, E.M., Stewart, V., Mendelsohn, M., Charron, J., Datta, M., Young, F., Stall, A.M. & Alt, F.W. 1992 RAG-2-deficient mice lack mature lymphocytes owing to inability to initiate V(D)J rearrangements. *Cell* **68**, 855–867.

Siegel, R.M., Katsumata, M., Miyashita, T., Louie, D.C., Greene, M.I. & Reed, J.C. 1992 Inhibition of thymocyte apoptosis and negative antigenic selection in *bcl-2* transgenic mice. *Proc. natn. Acad. Sci. U.S.A.* **89**, 7003–7007.

Smith, C.A., Williams, G.T., Kingston, R., Jenkinson, E.J. & Owen, J.J.T. 1989 Antibodies to CD3/T-cell receptor complex induce death by apoptosis in immature T cells in thymic cultures. *Nature, Lond.* **337**, 181–184.

Strasser, A., Harris, A.W., Bath, M.L. & Cory, S. 1990*a* Novel primitive lymphoid tumours induced in transgenic mice by cooperation between *myc* and *bcl-2*. *Nature, Lond.* **348**, 331–333.

Strasser, A., Harris, A.W. & Cory, S. 1991*a* Bcl-2 transgene inhibits T cell death and perturbs thymic self-censorship. *Cell* **67**, 889–899.

Strasser, A., Harris, A.W. & Cory, S. 1993 Eμ-*bcl-2* transgene facilitates spontaneous transformation of early pre-B and immunoglobulin-secreting cells but not T cells. *Oncogene* **8**, 1–9.

Strasser, A., Harris, A.W., Concoran, L.M. & Cory, S. 1994*a bcl-2* expression promotes B- but not T-lymphoid development in *scid* mice. *Nature, Lond.* **368**, 457–460.

Strasser, A. Harris, A.W., Jacks, T. & Cory, S. 1995 DNA damage can induce apoptosis in proliferating lymphoid cells via p53-independent mechanisms inhibitable by Bcl-2. (Submitted.)

Strasser, A., Harris, A.W., Vaux, D.L., Webb, E., Bath, M.L., Adams, J.M. & Cory, S. 1990*b* Abnormalities of the immune system induced by dysregulated *bcl-2* expression in transgenic mice. *Curr. Top. Microbiol. Immunol.* **166**, 175–181.

Strasser, A., Harris, A.W., von Boehmer, H. & Cory, S. 1994*b* Positive and negative selection of T cells in T-cell receptor transgenic mice expressing a *bcl-2* transgene. *Proc. natn. Acad. Sci. U.S.A.* **91**, 1376–1380.

Strasser, A., Whittingham, S., Vaux, D.L., Bath, M.L., Adams, J.M., Cory, S. & Harris, A.W. 1991*b* Enforced *BCL2* expression in B-lymphoid cells prolongs antibody responses and elicits autoimmune disease. *Proc. natn. Acad. Sci. U.S.A.* **88**, 8661–8665.

Tsujimoto, Y., Finger, L.R., Yunis, J., Nowell, P.C. & Croce, C.M. 1984 Cloning of the chromosome breakpoint of neoplastic B cells with the t(14;18) chromosome translocation. *Science, Wash.* **226**, 1097–1099.

Vaux, D.L., Cory, S. & Adams, J.M. 1988 *Bcl-2* gene promotes haemopoietic cell survival and cooperates with c-*myc* to immortalize pre-B cells. *Nature, Lond.* **335**, 440–442.

Vaux, D.L., Weissman, I.L. & Kim, S.K. 1992 Prevention of programmed cell death in *Caenorhabditis elegans* by human *bcl-2*. *Science, Wash.* **258**, 1955–1957.

von Boehmer, H. 1990 Developmental biology of T cells in T cell-receptor transgenic mice. *A. Rev. Immunol.* **8**, 531–556.

Williams, G.T., Smith, C.A., Spooncer, E., Dexter, T.M. & Taylor, D.R. 1990 Haemopoietic colony stimulating factors promote cell survival by suppressing apoptosis. *Nature, Lond.* **343**, 76–79.

Wyllie, A.H. 1980 Glucocorticoid-induced thymocyte apoptosis is associated with endogenous endonuclease activation. *Nature, Lond.* **284**, 555–556.

Wyllie, A.H., Kerr, J.F.R. & Currie, A.R. 1980 Cell death: the significance of apoptosis. *Int. Rev. Cytol.* **68**, 251–306.

Yunis, J.J., Frizzera, G., Oken, M.M., McKenna, J., Theologides, A. & Arnesen, M. 1987 Multiple recurrent genomic defects in follicular lymphoma. A possible model for cancer. *N. Engl. J. Med.* **316**, 79–84.

11

Molecular mechanisms for B lymphocyte selection: induction and regulation of antigen-receptor-mediated apoptosis of mature B cells in normal mice and their defect in autoimmunity-prone mice

TAKESHI TSUBATA, MASAO MURAKAMI, SAZUKU NISITANI AND TASUKU HONJO

Department of Medical Chemistry, Faculty of Medicine, Kyoto University, Kyoto 606, Japan

SUMMARY

Apoptosis (programmed cell death) has been suggested to be involved in clonal elimination of self-reactive lymphocytes for the normal function of the immune system. By crosslinking the antigen receptor (surface immunoglobulin; sIg) on the peritoneal B cells of normal mice, we found that strong crosslinking of sIg induces apoptosis of mature B cells, suggesting that interaction with membrane-bound self-antigens may eliminate self-reactive mature B cells by apoptosis. Antigen-receptor-mediated B cell apoptosis is blocked when a signal is transduced via the CD40 molecule on the B cell surface. Because the ligand of CD40 (CD40L) is expressed on activated T helper cells, B cells may escape from apoptosis and are activated when the immune system interacts with foreign antigens, which are normally able to activate T helper cells. Moreover, sIg crosslinking fails to induce apoptosis of both bcl-2-transgenic mice and autoimmune-disease-prone New Zealand mice. In these mice, the defect in sIg-mediated apoptosis of mature B cells may allow generation of self-reactive B cells, resulting in pathogenic consequences.

1. INTRODUCTION

Since Chiller *et al.* (1970) first demonstrated that tolerance can be induced within the B cell compartment, tolerance of self-reactive B cells has been suggested to be involved in the prevention of auto-immunity in the immune system. Self-tolerance of B cells may be maintained by elimination (clonal deletion) or functional inactivation of self-reactive B cells (clonal anergy) (Nossal 1983). Both of these mechanisms have been clearly demonstrated in autoantibody-transgenic mice, in which almost all the B cells express exogenous autoantibody genes and thus react to given self-antigens (Goodnow 1992). In autoantibody transgenic mice, self-reactive B cells are tolerated at the immature stage in the bone marrow. Immature B cells are thus likely to be the targets of the self-tolerance mechanisms: when self-reactive B cells in the bone marrow interact with the self-antigens, a signal via the antigen receptor (surface immunoglobulin; sIg) may either inactivate or eliminate self-reactive immature B cells. This assumption agrees with previous findings that immature B cells in the foetus, neonates or adult bone marrow are tolerated more easily than mature B cells in the peripheral lymphoid organs. Recent studies on auto-antibody-transgenic mice, however, have demon-strated that self-reactive B cells in the periphery are also tolerated upon interaction with self-antigens (Goodnow *et al.* 1989; Russel *et al.* 1991). This observation suggests that sIg can transduce a tolero-genic signal in mature B cells, although signalling through sIg most likely initiates an antibody response to foreign antigens by activating mature B cells. Here we demonstrate that signalling through sIg is able to induce apoptotic cell death of mature B cells in the peritoneal cavity and discuss the mechanisms determin-ing the fate (activation versus cell death) of mature B cells stimulated by antigens.

2. ANTIGEN-RECEPTOR-MEDIATED APOPTOSIS OF MATURE B CELLS

In mice transgenic for anti-red blood cell (RBC) autoantibody, numbers of B cells are markedly reduced in the bone marrow, spleen, lymph nodes and peripheral blood (Okamoto *et al.* 1992). This observation indicates that RBC-reactive B cells are deleted upon interaction with self-antigens (RBC). However, we found normal numbers of B cells in the peritoneal cavity of the transgenic mice. Interestingly, almost all of the peritoneal B cells in the transgenic mice are B-1 cells, which constitute a distinct B cell subset from conventional B cells, whereas the peritoneal cavity of normal mice contain both conventional B and B-1 cells. Several lines of evidence suggest that a small number of B-1 cells which, by a yet to be unidentified mechanism, have escaped from clonal deletion, migrate to the peritoneal cavity and extensively expand there (Murakami *et al.* 1992). Unlike conventional B cells, B-1 cells are able to expand in the peritoneal cavity because they possess self-replenishing capacity. Moreover, sequestration of peritoneal B cells from the RBC self-antigen seems to be essential for the expansion of self-reactive B-1 cells. Indeed, intraperitoneal injection of RBC results in massive apoptotic cell death of peritoneal B-1 cells in transgenic mice.

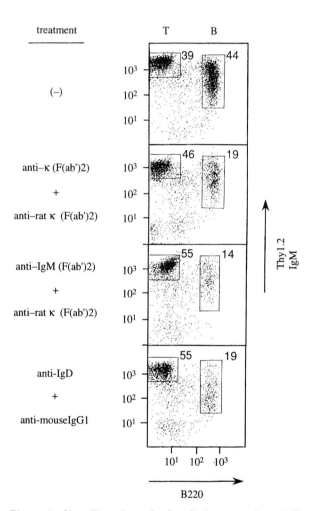

Figure 1. Signalling through sIg eliminates peritoneal B cells *in vivo*. C57BL/6 mice (6 week old) were injected intraperitoneally with 200 μg of the F(ab')$_2$ fragments of either R33-18 (rat anti-mouse κ) or Ak9 (rat anti-mouse IgM) followed by the injection of 200 μg of F(ab')$_2$ fragments of MAR18.5 (mouse anti-rat κ). Alternatively, C57BL/6 mice at the same age were injected intraperitoneally with the combination of 414/D7 (anti-mouse IgD^b, mouse IgG$_1$) and Ig(4a)10.9 (anti-mouse IgG$_1^a$). After 12 h, peritoneal cells were recovered, stained for Thy1, B220 and IgM, and analysed by two colour flow cytometry using a FACScan. Small lymphocytes were gated by forward versus side scatter. Percentages of B and T cell populations are indicated. Note that, by anti-Ig treatment, percentages of B cells are markedly reduced, whereas those of T cells are rather increased, suggesting that anti-Ig specifically eliminates B cells. Taken from Tsubata *et al.* (1994) with permission.

This finding indicates that signalling through the antigen receptor induces apoptosis of self-reactive mature B cells in these transgenic mice.

To determine whether signalling through sIg induces apoptosis of normal B cells, we injected anti-Ig antibodies into the peritoneal cavity of 6–8 week old normal C57BL/6 mice (Tsubata *et al.* 1994). We crosslinked sIg of peritoneal B cells by intraperitoneal injection of 200 μg each of anti-Ig and a second antibody which binds to the anti-Ig. Both B-1 cells and conventional B cells in the peritoneal cavity of the normal mice underwent apoptosis within 12 h (figures 1 and 2). As treatment with F(ab')$_2$ preparations of anti-Ig and the second antibody also induced apoptosis of peritoneal B cells, this reaction does not require signalling through the Fcγ receptor, which has been shown to transduce a negative signal in B cells. In contrast, injection of smaller amount of F(ab')$_2$ preparation of anti-Ig alone does not induce apoptosis but enhances class II MHC expression of peritoneal B cells, suggesting that this treatment induces signalling through sIg for B cell activation. This result agrees with previous findings that anti-Ig treatment activates mature B cells in both *in vitro* (Parker 1980; DeFranco *et al.* 1982) and *in vivo* systems (Finkelman *et al.* 1982), although proliferation and differentiation of the activated B cells require T cell-derived cytokines. Taken together, strong sIg crosslinking eliminates normal mature B cells by apoptosis, whereas weaker sIg crosslinking activates mature B cells. It is, however, not yet clear whether mature B cells in lymphoid organs other than the peritoneal cavity undergo apoptosis by sIg crosslinking. Because membrane-bound antigens crosslink sIg more strongly than soluble antigens, interaction with membrane-bound antigens may induce apoptosis of normal mature B cells as well as B cells in anti-RBC transgenic mice. Soluble antigens, in contrast, may activate responding B cells.

3. REGULATION OF ANTIGEN-RECEPTOR-MEDIATED APOPTOSIS OF B LYMPHOCYTES

The finding that strong sIg crosslinking induces apoptosis of mature B cells made us ask how the immune system responds to membrane-bound foreign antigens such as antigens on the bacterial wall. We examined the molecular mechanisms that inhibit antigen-receptor-mediated B cell apoptosis.

The bcl-2 proto-oncogene was originally identified because of its frequent translocation in follicular lymphomas (Bakhshi *et al.* 1985; Cleary & Sklar 1985; Tsujimoto *et al.* 1985). Recent studies have revealed that expression of bcl-2 rescues both hemopoietic cells and nervous cells from apoptosis in many, but not all, experimental systems (Vaux *et al.* 1988; Hockenberry *et al.* 1990; Nunez *et al.* 1990; Garcia *et al.* 1992; Allsopp *et al.* 1993). We thus injected anti-Ig and the second antibody into the peritoneal cavity of bcl-2-transgenic mice, which produce a large amount of bcl-2 in B cells. Strong sIg crosslinking does not induce apoptosis but activates peritoneal B cells of bcl-2 transgenic mice (Tsubata *et al.* 1994). This observation suggests

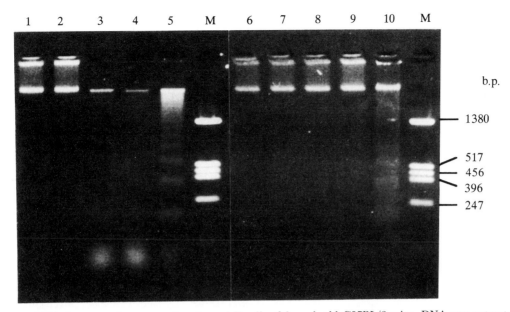

Figure 2. DNA fragmentation assay of peritoneal B cells of 6 week old C57BL/6 mice. DNA was extracted from the peritoneal cells of non-treated mice (lanes 1 and 6), mice treated with rat anti-mouse κ (187.1) (lanes 3 and 8) or with anti-rat IgG (purchased from Zymed) (lanes 4 and 9), mice intraperitoneally injected with both rat anti-mouse B220 (RA3-6B2) and anti-rat IgG (lanes 2 and 7) and mice treated with both anti-κ and anti-rat IgG (lanes 5 and 10). Treatment of mice was done as indicated in the legend to figure 1. Low molecular mass genomic DNA (lanes 1–5) and high molecular mass DNA (lane 6–10) was prepared and analysed by agarose gel electrophoresis. Molecular mass markers (M) in base pairs (b.p.) are indicated. Note that treatment with the combination of anti-κ and the second antibody (anti-rat IgG) induced DNA fragmentation, a marker for apoptosis, whereas the treatment with either anti-κ alone, the second antibody alone or the combination of the control antibody (anti-B220) and the second antibody failed. Taken from Tsubata *et al.* (1994) with permission.

that, in the presence of rescue molecules like bcl-2, B cells are able to respond to membrane-bound antigens.

As is the case for normal peritoneal B cells, WEHI-231 B lymphoma cells undergo apoptosis upon sIg crosslinking by anti-Ig (Benhamou *et al.* 1990; Hasbold & Klaus 1990). Scott *et al.* (1987) demonstrated that in the presence of T helper cells, WEHI-231 cells are resistant to anti-Ig treatment. T helper cells may express the molecule(s) which induce(s) B cells to express molecules protecting the cells from apoptosis. To determine the molecules involved in the rescue of WEHI-231 cells from apoptosis, we compared T helper lines capable of blocking apoptosis of WEHI-231 cells and those incapable (Tsubata *et al.* 1993). We found that T helper lines having the rescuing ability of WEHI-231 cells from apoptosis expressed the ligand for the B cell antigen CD40 (CD40L) (Clark & Ledbetter 1986; Stamenkovic *et al.* 1989; Armitage *et al.* 1992); those without the rescuing ability did not. We then demonstrated directly the involvement of CD40-mediated signalling in rescuing anti-Ig-treated WEHI-231 cells by treating WEHI-231 cells with the anti-CD40 antibody, soluble chimeric molecules containing the active portion of CD40L or CD40L-transfected cells. Any of these treatments protected WEHI-231 cells from antigen-receptor-mediated apoptosis. Furthermore, WEHI-231 cells upregulate class II MHC expression when treated with both anti-Ig and anti-CD40, indicating that B cells are activated in the co-existence of antigens and T-helper-cell-derived signals. Taken together, B

cells may undergo apoptosis upon interaction with membrane-bound antigens alone, and require a T-helper-cell-derived rescue signal for responding to the membrane-bound antigens. This model is in agreement with the two-signal model for B cell activation originally proposed by Bretcher & Cohn (1970). As CD40L is expressed on activated T helper cells but not resting T cells (Lane *et al.* 1992; Noelle *et al.* 1992), B cells are most likely rescued when the antigens are able to stimulate T cells as well as B cells. Antigen-receptor–mediated apoptosis may thus eliminate self-reactive B cells generated within the mature B cell compartment because self-reactive T cells are eliminated in the thymus. It is of note that mature T cells do not change their antigen specificity because they do not undergo somatic mutation of the antigen receptor genes. In contrast, membrane-bound foreign antigens stimulate both T and B cells, resulting in rescuing antigen-stimulated B cells from apoptosis by signalling through CD40.

4. DEFECT IN ANTIGEN-RECEPTOR-MEDIATED APOPTOSIS IN AUTOIMMUNITY

Although self-reactive B cells may be tolerated at the immature B cell stage, self-tolerance mechanisms may also be required at the mature B cell stage. Indeed, self-reactive B cells are generated within the mature B cell compartment from B cells that have no self-reactivity by somatic mutations of Ig V genes (Diamond & Scharff 1984; Giusti *et al.* 1987). Furthermore, self-reactive B cells reacting to the self-antigens

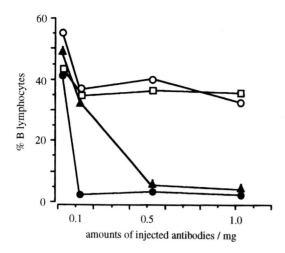

Figure 3. Failure of sIg-mediated apoptosis of peritoneal B cells in autoimmunity-prone NZB and (NZB × NZW)F1 mice. Autoimmunity-prone NZB (open circles) and (NZB × NZW)F1 (open squares) mice as well as normal C57BL/6 (filled circles) and NZW (filled triangles) mice (6 week old) were injected intraperitoneally with 100, 500 or 1000 μg of MB86 (anti-mouse IgM^b) and the same amount of the second antibody (goat anti-mouse IgG1). After 12 h, the peritoneal cells were collected and stained for B220 and IgM. Percentages of B cells (B220$^+$, IgM$^+$) were determined by flow cytometry as described in the legend to figure 1. Taken from Tsubata *et al.* (1994) with permission.

localized in the periphery are not eliminated at the immature stage in the bone marrow. Those self-reactive mature B cells may be eliminated upon interaction with self-antigens. Defects in this mechanism may thus allow self-reactive B cells to survive, resulting in autoantibody production.

Anti-DNA antibodies are characteristic of systemic lupus erythematosus (SLE), a prototypic autoantibody-mediated autoimmune disease. B cells reactive to DNA have been shown to be either eliminated or functionally inactivated in the anti-DNA-transgenic mice of the normal background, suggesting that anti-DNA B cells are tolerated in normal mice. In contrast, a large amount of anti-DNA antibodies is produced in bcl-2-transgenic mice, which also develop SLE-like immune complex nephritis (Strasser *et al.* 1991). This finding suggests some defect in the self-tolerance mechanisms in the bcl-2-transgenic mice. Indeed, peritoneal B cells do not undergo apoptosis upon anti-Ig injection, suggesting that bcl-2 transgenic mice fail to eliminate self-reactive mature B cells (figure 4). However, Goodnow's group (Hartley *et al.* 1993) and ours (Nisitani *et al.* 1993) have shown that bcl-2 does not block clonal deletion of self-reactive B cells at the immature stage in the bone marrow. Self-reactive B cells may thus emerge within the mature B cell compartment in the bcl-2 transgenic mice presumably due to the defect in antigen-receptor-mediated apoptosis. This finding also suggests the importance of self-tolerance mechanisms in the mature B cell compartment in protecting normal individuals from autoimmunity.

F1 hybrids of NZB and NZW (BWF1) mice produce anti-DNA antibody and develop nephritis spontaneously. NZB mice develop autoimmune hemolytic anemia by producing anti-RBC autoantibodies. To examine the B cell tolerance mechanisms of these autoimmune-disease-prone mice, we treated 6–8 week old NZB and BWF1 mice with intraperitoneal injection of anti-Ig, followed by the injection of the second antibody (Tsubata *et al.* 1994). In both strains of mice, percentages of B cells in the peritoneal cells are only slightly reduced even by the injection of 1 mg each of anti-Ig and the second antibody (figure 3). In contrast, almost all the B cells undergo apoptosis by the injection of 200 μg each of anti-Ig and the second antibody in normal C57BL/6 mice. Slight reduction in

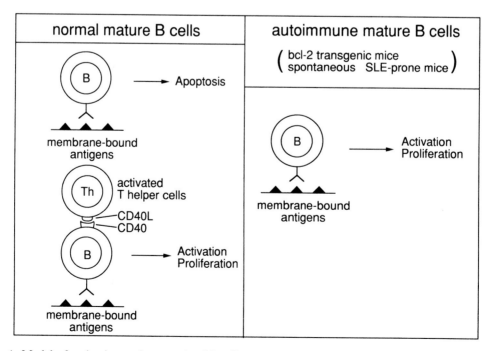

Figure 4. Model of activation and apoptosis of B cells upon interaction with membrane-bound antigens. Taken from Tsubata *et al.* (1994) with permission.

B cell percentage in the autoimmunity-prone mice may be due to the expansion of non-lymphocytic inflammatory cells by anti-Ig injection. Mature B cells of the autoimmunity-prone mice are likely to be resistant to antigen-receptor-mediated apoptosis and to allow self-reactive B cells to survive (figure 4). It has been demonstrated that high-affinity anti-DNA antibodies, which are presumably pathogenic, are produced by extensive somatic mutations of Ig V genes (Hirose *et al.* 1993). This result suggests that those pathogenic autoantibodies are generated in the mature B cell compartment as is the case for bcl-2-transgenic mice. Although, we failed to detect enhanced expression of bcl-2 in the autoimmunity-prone mice, those mice may produce an excess amount of bcl-2-related molecules such as bcl-x (Boise *et al.* 1993) or Mcl-1 (Kozopas *et al.* 1993). Molecular mechanisms responsible for the resistance of NZB and BWF1 B cells to apoptosis are now under investigation.

REFERENCES

Allsopp, T.E., Wyatt, S., Paterson, H.F. & Davies, A.M. 1993 The proto-oncogene bcl-2 can selectively rescue neurotrophic factor-dependent neurons from apoptosis. *Cell* **73**, 295–307.

Bakhshi, A., Jensen, J.P., Goldman, P., Wright, J.J., McBride, O.W., Epstein, A.L. & Korsmeyer, S.J. 1985 Cloning the chromosomal breakpoint of t(14;18) human lymphomas: clustering around J_H on chromosome 14 and near a transcriptional unit on 18. *Cell* **41**, 899–906.

Benhamou, L.E., Cazenave, P.-A. & Sarthou, P. 1990. Anti-immunoglobulins induce death by apoptosis in WEHI-231 B lymphoma cells. *Eur. J. Immunol.* **20**, 1405–1407.

Bretscher, P. & Cohn, M. 1970 A theory of self-nonself discrimination: paralysis and induction involve the recognition of one and two determinants on an antigen, respectively. *Science, Wash.* **169**, 1042–1049.

Chiller, J.M., Habicht, G.S. & Weigle, W.O. 1970 Cellular sites of immunologic unresponsiveness. *Proc. natn. Acad. Sci. U.S.A.* **65**, 551–556.

Cleary, M.L. & Sklar, J. 1985 Nucleotide sequence of a t(14;18) chromosomal breakpoint in follicular lymphoma and demonstration of a breakpoint-cluster region near a transcriptionally active locus on chromosome 18. *Proc. natn. Acad. Sci. U.S.A.* **82**, 7439–7443.

DeFranco, A.L., Kung, J.T. & Paul, W.E. 1982 Regulation of growth and proliferation in B cell subpopulations. *Immunol. Rev.* **64**, 161–182.

Finkelman, F.D., Scher, I., Mond, J.J., Kung, J.T. & Metcalf, E.S. 1982 Polyclonal activation of the murine immune system by an antibody to IgD. I. Increase in cell size and DNA synthesis. *J. Immunol.* **129**, 629–637.

Garcia, I., Martinou, I., Tsujimoto, Y. & Martinou, J.-C. 1992 Prevention of programmed cell death of sympathetic neurons by the *bcl-2* proto-oncogene. *Science, Wash.* **258**, 302–304.

Goodnow, C.C., Crosbie, J., Jorgensen, H., Brink, R.A. & Basten, A. 1989 Induction of self-tolerance in mature peripheral B lymphocytes. *Nature, Lond.* **342**, 385–391.

Goodnow, C.C. 1992 Transgenic mice and analysis of B-cell tolerance. *A. Rev. Immunol.* **10**, 489–518.

Hartley, S.B., Cooke, M.P., Fulcher, D.A. *et al.* 1993 Elimination of self-reactive B lymphocytes proceeds in two stages: arrested development and cell death. *Cell* **72**, 325–335.

Hasbold, J. & Klaus, G.G.B. 1990 Anti-immunoglobulin antibodies induce apoptosis in immature B cell lymphomas. *Eur. J. Immunol.* **20**, 1685–1690.

Hirose, S., Wakiya, M., Kawano-Nishi, Y. *et al.* 1993 Somatic diversification and affinity maturation of IgM and IgG anti-DNA antibodies in murine lupus. *Eur. J. Immunol.* **23**, 2813–2820.

Hockenberry, D., Nuñez, G., Milliman, C., Schreiber, R.D. & Korsmeyer, S.J. 1990 Bcl-2 is an inner mitochondrial membrane protein that blocks programmed cell death. *Nature, Lond.* **348**, 334–339.

Murakami, M., Tsubata, T., Okamoto, M. *et al.* 1992 Antigen-induced apoptotic death of Ly-1 B cells responsible for autoimmune disease in transgenic mice. *Nature, Lond.* **357**, 77–80.

Nisitani, S., Tsubata, T., Murakami, M., Okamoto, M. & Honjo, T. 1993 The bcl-2 gene product inhibits clonal deletion of self-reactive B lymphocytes in the periphery but not in the bone marrow. *J. exp. Med.* **178**, 1247–1254.

Nossal, G.J.V. 1983 Cellular mechanisms of immunologic tolerance. *A. Rev. Immunol.* **1**, 33–62.

Nuñez, G., London, L., Hockenberry, D., Alexander, M., McKearn, J.P. & Korsmeyer, S.J. 1990 Deregulated Bcl-2 gene expression selectively prolongs survival of growth factor-deprived hemopoietic cell lines. *J. Immunol.* **144**, 3602–3610.

Okamoto, M., Murakami, M., Shimizu, A. *et al.* 1992 A transgenic model of autoimmune hemolytic anemia. *J. exp. Med.* **175**, 71–79.

Parker, D.C. 1980 Induction and suppression of polyclonal antibody responses by anti-Ig reagents and antigen-nonspecific helper factors. *Immunol. Rev.* **52**, 115–139.

Russel, D.M., Dembić, Z., Morahan, G., Miller, J.F.A.P., Bürki, K. & Nemazee, D. 1991 Peripheral deletion of self-reactive B cells. *Nature, Lond.* **354**, 308–311.

Scott, D.W., O'Garra, A., Warren, D. & Klaus, G.G.B. 1987 Lymphoma models for B cell activation and tolerance. VI. Reversal of anti-Ig-mediated negative signalling by T cell-derived lymphokines. *J. Immunol.* **139**, 3924–3929.

Strasser, A., Whittingham, S., Vaux, D.L. *et al* 1991 Enforced *BCL2* expression in B-lymphoid cells prolongs antibody responses and elicits autoimmune disease. *Proc. natn. Acad. Sci. U.S.A.* **88**, 8661–8665.

Tsubata, T., Wu, J. & Honjo, T. 1993 B-cell apoptosis induced by antigen receptor crosslinking is blocked by T-cell signal through CD40. *Nature, Lond.* **364**, 645–648.

Tsubata, T., Murakami, M. & Honjo, T. 1994 Antigen-receptor cross-linking induces peritoneal B-cell apoptosis in normal but not autoimmunity-prone mice. *Curr. Biol.* **4**, 8–17.

Tsujimoto, Y., Cossman, J., Jaffe, E. & Croce, C.M. 1985 Involvement of the *bcl-2* gene in human follicular lymphoma. *Science, Wash.* **228**, 1440–1443.

Vaux, D.L., Cory, S. & Adams, J.M. 1988 Bcl-2 gene promotes haemopoietic cell survival and cooperates with c-*myc* to immortalize pre-B cells. *Nature, Lond.* **335**, 440–442.

12

Fas-based d10S-mediated cytotoxicity requires macromolecular synthesis for effector cell activation but not for target cell death

M.-F. LUCIANI AND P. GOLSTEIN

Centre d'Immunologie INSERM-CNRS de Marseille-Luminy, Case 906, 13288 Marseille Cedex 9, France

SUMMARY

Two main mechanisms seem at play in T cell-mediated cytotoxicity, a process in which target cell death often follows an apoptotic cell death pattern. One of these involves Fas at the target cell surface and a Fas ligand at the effector cell surface. This allowed us to reinvestigate the long-standing question of macromolecular synthesis requirement in T cell-mediated cytotoxicity, using the d10S model cell line which is cytotoxic apparently only via the Fas molecularly defined mechanism. We showed, first, that induction of cytotoxic activity of effector cells, obtained by preincubating these effector cells with a phorbol ester and a calcium ionophore, could be inhibited by macromolecular synthesis inhibitors (cycloheximide, actinomycin D, DRB). We then investigated whether macromolecular synthesis was required, when effector and target cells were mixed, to obtain target cell death. Preincubating already activated effector cells for 30 min with macromolecular synthesis inhibitors, then adding target cells and performing the ^{51}Cr release cytotoxicity test in the presence of these inhibitors, did not significantly decrease subsequent target cell death, indicating that already activated effector cells could kill without further requirement for macromolecular synthesis. In addition, target cell preincubation for up to 3 h in the presence of one of these inhibitors did not decrease cell death. The high sensitivity of mouse thymocytes to this type of cytotoxicity enabled us to devise the following experiment. As previously shown by others, thymocyte death induced by dexamethasone (DEX) could be blocked by coincubation with cycloheximide (CHX). Such DEX-treated CHX-rescued thymocytes, the survival of which was an internal control of efficiency of protein synthesis inhibition, were then subjected to effector cells in the presence of CHX, and were shown to die. Thus, there is no requirement for macromolecular synthesis at the target cell level in this variety of apoptotic cell death. Altogether, in this defined mechanism of T cell-mediated cytotoxicity, macromolecular synthesis is required for d10S effector cell activation, but not for lysis by already activated effector cells nor for target cell death.

1. INTRODUCTION

The main criteria of apoptosis (Kerr *et al.* 1972), one of several ways used by a cell to die, are morphological alterations such as cytoplasmic and nuclear condensation/fragmentation, and often DNA fragmentation and a requirement for macromolecular synthesis (MMS) (reviewed by Cohen 1991; Ellis *et al.* 1991; Golstein *et al.* 1991; Kerr & Harmon 1991; Lockshin & Zakeri 1991). MMS in the context of this report should be regarded as mRNA and protein synthesis. Several distinct types of cell death may exist under the collective concept of apoptosis. They might be distinguished according to their faithfulness to the above-mentioned criteria (see discussion in Golstein *et al.* 1991). Thus, sometimes a morphologically apoptotic cell death does not require MMS (see below), or is facilitated if MMS is inhibited (Ruff & Gifford 1981), or can even be triggered by MMS inhibitors (Searle *et al.* 1975; Martin *et al.* 1990; Cohen 1991). The target

cell death induced by cytotoxic T cells (Berke 1989; Tschopp & Nabholz 1990; Young *et al.* 1990; Duke 1991; Golstein *et al.* 1991; Podack *et al.* 1991), considered apoptotic on the basis of morphologic criteria and DNA fragmentation, was reported not to require MMS (Duke *et al.* 1983). More recent results, however, suggested that target cell MMS may be required in certain cases or circumstances of T cell-mediated cytotoxicity (Landon *et al.* 1990; Zychlinsky *et al.* 1991). These apparent discrepancies might be due, in particular, to the fact that different mechanisms of T cell-mediated cytotoxicity (Duke 1991) were considered.

Recently, we showed (Rouvier *et al.* 1993) that most if not all of the calcium-independent component of T cell-mediated cytotoxicity involved the death-transducing Fas molecule (Yonehara *et al.* 1989; Trauth *et al.* 1989) at the target cell surface. This Fas-based mechanism was first detected using d10S model cytotoxic cells (Rouvier *et al.* 1993), but was not limited to these. It could be shown to account, in particular, for the calcium-independent component of cytotoxicity by populations of alloimmune peritoneal exudate lymphocytes (Rouvier *et al.* 1993) and of mixed leukocyte culture cells, and by several cytotoxic T cell clones (M.-F. Luciani & P. Golstein, unpublished observations). This mechanism is defined in molecular terms by a requirement for Fas at the target cell surface (Rouvier *et al.* 1993) and by a requirement for the Fas ligand (Suda *et al.* 1993) at the effector cell surface. In parallel, a perforin-based mechanism, detected some years ago (Henkart 1985; Podack 1985), was formally demonstrated as a major mechanism of T cell-mediated cytotoxicity through the use of perforin knock-out mice (D. Kägi & H. Hengartner, personal communication).

In this study, we reinvestigate the requirement for MMS in the Fas-based molecularly defined system of T cell-mediated cytotoxicity. We took advantage of the availability of prototypic effector cells (d10S) exerting cytotoxicity apparently only via the Fas-dependent mechanism, and of the sensitivity of thymocytes in this system, allowing their use in cell-death-inhibition internally controlled experiments. We thus demonstrated, first, that there is MMS requirement for d10S effector cell activation, and, second, that there is no significant requirement for MMS at the actual killing stage, and especially in dying target cells, in this mechanism of T cell-mediated cell death.

2. MATERIALS AND METHODS

(a) Culture conditions, cells and reagents

All incubation and culture procedures were done at 37°C in a water-saturated 7% CO_2 atmosphere, in RPMI 1640 or DME medium (Gibco Bio-Cult, Glasgow, U.K.) enriched with 5% foetal calf serum (FCS, Biological Industries, Israel). Target cells were either YAC tumour cells or thymocytes freshly explanted from 6–8 week-old C57Bl/6 mice.

PC60 cells (Conzelmann *et al.* 1982), a hybridoma between a mouse cytotoxic T cell clone with anti-male D^b specificity (Von Boehmer *et al.* 1979) and a derivative from the rat T lymphoma W/Fu (C58NT)D, and all PC60-derived (PC60-d) cells including PC60-d10S cells (d10S for short) were grown and cloned in Dulbecco medium (Gibco Biocult) enriched with 5% FCS. Cloning was done by limiting dilution in flat-bottomed wells of 96-well tissue culture microplates (C.E.B., France). Most of the PC60-d clones were derived from wells having received an average number of 0.3 cells per well. After sufficient growth, cloned cells were transferred to wells of 24-well tissue culture plates (Costar, Cambridge, Massachusetts) and later to tissue culture flasks (Falcon, Becton Dickinson, Lincoln Park, New Jersey). These PC60-d cloned cells were then induced to check their cytotoxic potential. Induction of cytolysis was by addition, at the beginning of the cytotoxicity test or in preincubation experiments, of a mixture of phorbol myristic acetate (PMA; Sigma, Saint-Louis, Missouri; final concentration 10 ng ml^{-1})

and of the Ca^{2+} ionophore ionomycin (Calbiochem, San Diego, California; final concentration 3 µg ml^{-1}). d10S cells activated by preincubation for 3 h with PMA and ionomycin are denoted d10S PI. The protein synthesis inhibitor cycloheximide (CHX) and the RNA synthesis inhibitors actinomycin D and dichlororibofuranosylbenzimidazole (DRB) were all from Sigma.

(b) Cell-mediated cytotoxicity tests

^{51}Cr release tests were carried out in V-shaped wells of 96-well microtitre plates with ^{51}Cr-labelled target cells (either 10^4 YAC or 10^5 thymocytes) and effector cells at the indicated ratios in a total volume per well of 200 µl of RPMI medium with 10% FCS. The plates were centrifuged (200 *g*, 2 min) and incubated for 4 h unless specified otherwise. After another centrifugation, 100 µl aliquots of supernates were assayed for radioactivity. The fraction of the total radioactivity released was then calculated, and the results were expressed as % experimental ^{51}Cr release minus % ^{51}Cr release from target cells alone, except for the experiment in figure 3*d*.

Some experiments with thymocytes (figure 3*a–c*) were assayed using a trypan blue exclusion test. Somewhat erratic results in preliminary experiments were found to be due to variations of thymocyte staining with time after addition of trypan blue to the thymocyte suspensions. In the experiments presented in this report, trypan blue was added at the indicated times to the cell suspensions, which were kept for a further period of time of 20 to 40 min at room temperature and only then microscopically assessed for cell death; this resulted in the presence of two unambiguously distinct types of cells, 'white' with contrasted edges considered living, and blue at various stages of disintegration. Because the number of cells having died in a culture may exceed the number of remaining blue cells, only living cells were counted. The results are expressed as % surviving cells (concentration of living cells in experimental group/initial concentration of living cells ×100).

3. RESULTS

(a) Preliminary considerations

The inducible cytotoxic T cell hybridoma PC60 (Conzelmann *et al.* 1982), obtained from M. Nabholz, was serially subcloned, with systematic passaging of the clones that after induction were most cytotoxic against ^{51}Cr-labelled YAC target cells. A clone obtained after the tenth serial subcloning, called d10S, was used for the experiments described in this report. After induction with a mixture of PMA and ionomycin, cytotoxicity by d10S cells has many characteristics of a calcium-independent MHC-unrestricted T cell cytotoxicity, and leads to DNA fragmentation in YAC target cells (not shown). d10S PI cells efficiently kill both YAC target cells and thymocytes, the latter at very low ratios of effector to target cells ((Rouvier *et al.* 1993); table 1).

Table 1. *Inhibition by cycloheximide of d10S effector cell activation by a mixture of PMA and ionomycin*

cytotoxicity test[b]			cytotoxic activity of d10S cells preincubated with PI[a]					
			without CHX			with CHX		
targets	CHX in test	E : T	3	1	0.3	3	1	0.3
YAC	−		79[c]	68	26	33	36	9
	+		77	51	22	11	5	5
		E : T	0.3	0.1	0.03	0.3	0.1	0.03
thymocytes	−		32	31	15	19	18	13
	+		46	21	9	14	8	3

[a] Effector d10S cells were preincubated for 3 h in the presence of PMA and ionomycin (PI) either with or without cycloheximide (CHX) at a final concentration of $10 \,\mu g \, ml^{-1}$, and washed by centrifugation.
[b] The 4 h cytotoxicity test was run at the indicated E : T (effector : target cell) ratio, in the presence of PI, and with or without cycloheximide.
[c] Results expressed as % experimental ^{51}Cr release minus % ^{51}Cr release of target cells alone (7–8 for YAC, 14–16 for thymocytes).

Importantly, all of the detectable cytotoxicity of d10S PI, at least when tested against thymocytes and against Fas-transfected L1210 tumour cells, requires the presence of Fas on the target cell surface (Rouvier *et al.* 1993); this is most probably also true for the equally calcium-independent (Rouvier *et al.* 1993) cytotoxicity of d10S PI tested against YAC tumour target cells. Using d10S PI as effector cells, in particular with thymocytes as target cells, seems therefore to be the best means towards exploring an isolated molecular mechanism of cell-mediated cytotoxicity.

(b) *Induction of d10S cytolytic activity by PMA and ionomycin requires macromolecular synthesis*

Table 1 shows that d10S activation by incubation for 3 h with PI before the cytotoxicity test led to less cytotoxicity if this preincubation was performed in the presence of CHX. Because of the reversibility of the CHX effect, the inhibition of induction of cytolysis was more marked if CHX was present, not only during preincubation, but also during the cytotoxicity test itself (table 1). d10S activation by PI was significantly inhibited by CHX at concentrations as low as $0.1 \,\mu g \, ml^{-1}$ (figure 1a, dotted line), and also by actinomycin D and DRB (figure 1b,c, dotted lines). The inhibition of activation was most probably related to inhibition of MMS, as it occurred using each of three drugs known to interfere with distinct steps of MMS, and at low concentrations of these drugs, in particular of CHX. d10S activation with PI thus most probably required MMS. Interestingly, after 3 h in PI the effector cells had similar cytolytic activity whether the cytotoxicity test was performed in the presence or in the absence of CHX (table 1), i.e. the PI-activated effector cells had synthesized by then all the macromolecules required to kill.

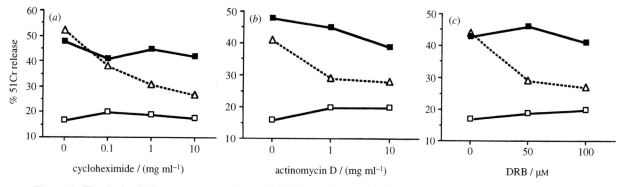

Figure 1. The lack of effect on cytotoxicity of inhibitors of MMS added at the beginning of the cytotoxicity test. In three separate experiments, cycloheximide (*a*), actinomycin D (*b*) or DRB (*c*) at the indicated final concentrations were added to d10S effector cells, preactivated (filled squares) or not (open squares) with a mixture of PMA and ionomycin. After 30 min of incubation at 37°C, YAC target cells were added and the cytotoxicity test proceeded for 4 h. The effector : target cell ratio was 1 : 1 in all cases. As a positive drug efficiency control, the same concentrations of inhibitors were used with non-preactivated d10S cells, and PMA and ionomycin were added directly to the cytotoxicity test; under these conditions the inhibitors affected activation, resulting in less cytotoxic activity generated during the test (dotted line). The cytotoxicity is expressed as % ^{51}Cr release as a function of inhibitor concentration. Spontaneous ^{51}Cr release was 10–20%.

(c) Lysis of target cells by PI-activated d10S effector cells requires macromolecular synthesis neither at the effector nor at the target cell level

In the experiments described in figure 1, already activated effector cells (d10S PI) were incubated for 30 min in the presence of the inhibitors before the beginning of the cytotoxicity test, which was performed also in the presence of these inhibitors. As in the experiment described in table 1, this led to little or no significant decrease in cytotoxicity, indicating that already activated effector cells did not require MMS to lyse target cells. Because target cells were lysed to approximatley the same degree, whether or not CHX was present in the cytotoxicity test, the experiments described in table 1 and in figure 1a also suggested that the target cells possessed all the proteins required to die. The lack of any significant decrease of cytotoxicity when either actinomycin D or DRB were added at the beginning of the cytotoxicity test (figure 1b,c) suggested that no RNA synthesis was required at the dying target cell level. These conclusions still applied when YAC or thymocyte target cells were preincubated for 30 min in the presence of CHX or of actinomycin D before the cytotoxicity test (figure 2). In fact, preincubation of YAC target cells with actinomycin D for as long as 3 h before the cytotoxicity test did not significantly decrease their sensitivity to d10S PI (not shown).

Although many of the experiments above were controlled in terms of drug activity (as the drugs were shown in the same experiments to inhibit effector d10S cell activation), they were not, however, internally controlled for efficiency of macromolecular synthesis inhibition at the target cell level. To this end, we took advantage of the observation that effector d10S PI cells could lyse not only YAC target cells, but also thymocytes, the latter moreover at very low ratios of effector to target cells (Rouvier *et al.* 1993; table 1, figure 2b,c). Thymocytes could also

be lysed by DEX in a classical *in vitro* apoptosis system (Wyllie 1980).

As shown before (Cohen & Duke 1984; Wyllie *et al.* 1984), thymocytes undergoing apoptotic death after 10–20 h of incubation with DEX (figure 3a) did not die if this incubation included CHX (figure 3b), showing that protein synthesis by the dying cells is required for death to occur in this case. Thymocytes treated with DEX and protected by addition of CHX were processed after overnight incubation in three different ways. When washed and resuspended in medium without CHX, they began to die, thus reflecting a long-term death signal given by DEX. This was shown in both a trypan blue viability assay (figure 3c) and a ^{51}Cr release assay (figure 3d). When resuspended in the presence of CHX, they did not die (figure 3c,d), thus reflecting the MMS requirement of DEX-induced apoptotic death and the efficiency of CHX to block this MMS. When d10S PI effector cells were added to these CHX-blocked thymocytes, the thymocytes died (figure 3d). The latter results, obtained in each of three similar experiments, showed that thymocytes can be killed by d10S PI effector cells even when the protein molecules whose synthesis is required for DEX-induced apoptosis of the same thymocytes are demonstrably not synthesized. A reservation to these conclusions would stem from the possibility, which we do not consider very likely, that effector cells would lyse with similar efficiency, but through a different mechanism, untreated thymocytes and thymocytes treated with DEX and CHX.

4. DISCUSSION

Studies on MMS requirements of T cell-mediated cytotoxicity, undertaken to provide indications as to the mechanism(s) of the latter, have included both effector and target cell preincubation or coincubation

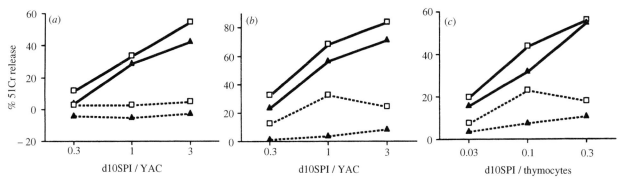

Figure 2. The lack of effect on cytotoxicity of target cell preincubation with inhibitors of MMS. In a first experiment (a), YAC target cells were incubated for 30 min at 37°C before the cytotoxicity test with medium alone (open squares) or with actinomycin D at a final concentration of $5\,\mu\mathrm{g\,ml}^{-1}$ (filled triangles). In a second experiment, YAC cells (b) or thymocytes (c) were incubated similarly with medium alone (open squares) or with cycloheximide at a final concentration of $10\,\mu\mathrm{g\,ml}^{-1}$ (filled triangles). Target cells preincubated with a drug were washed (to eliminate released ^{51}Cr) and were re-exposed to the same drug at the same final concentration during the 4 h cytotoxicity test. Effector cells were d10S-preactivated by incubation with PMA and ionomycin. As a positive drug efficiency control, this preactivation was performed in the presence of the corresponding drug (dotted lines) or in its absence (full lines). The cytotoxicity is expressed as % ^{51}Cr release as a function of effector to target cell ratio. Spontaneous ^{51}Cr release was 7–17%.

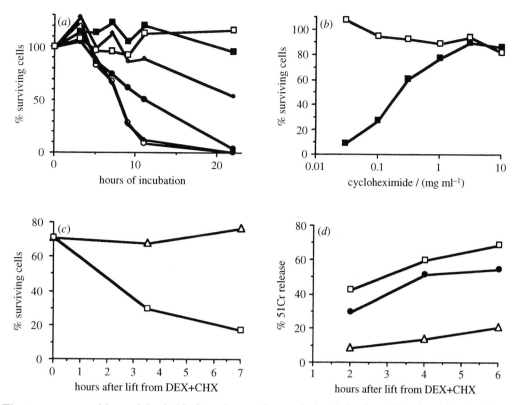

Figure 3. Thymocytes rescued by cycloheximide from dexamethasone-induced death could be lysed by d10S PI cells. (a) % surviving thymocytes after incubation for various lengths of time with either medium alone (open squares), or tenfold increasing concentrations of dexamethasone up to 10^{-6} M (open circles). (b) % surviving thymocytes after overnight incubation with medium (open squares) or 10^{-6} M dexamethasone (filled squares) and various concentrations of cycloheximide. Thymocytes thus incubated overnight with both dexamethasone (10^{-6} M) and cycloheximide ($10\,\mu g\,ml^{-1}$) were washed and reincubated either in medium alone (open squares) or with cycloheximide ($10\,\mu g\,ml^{-1}$; open triangles); death of cycloheximide-unprotected, but not of cycloheximide-protected thymocytes, was followed with time, in separate experiments, either as % surviving cells (c) or in a ^{51}Cr release test (d). Addition of d10S PI to cycloheximide-protected thymocytes, at a ratio of $0.1:1$, resulted in significant lysis of these thymocytes (d, filled circles). Results are given as uncorrected % ^{51}Cr release.

with inhibitors of MMS (Brunner *et al.* 1968; Thorn & Henney 1976; Landon *et al.* 1990; Zychlinsky *et al.* 1991). The variability of the results thus obtained may have originated from the use of different effector cells, different stages of activation of these, different target cells, different concentrations of inhibitors, or from the initially unsuspected heterogeneity of the mechanisms of T cell-mediated cytotoxicity under investigation. The availability of a T cell-mediated cytotoxicity system lysing via a defined mechanism through target-cell-surface Fas (Rouvier *et al.* 1993) enabled us to re-evaluate in this system the MMS requirements (and may be of use to investigate other parameters of T cell-mediated cytotoxicity). We reached two conclusions: in our d10S cell model system, MMS is required for the induction of cytotoxic activity, and is not required at the actual killing stage nor for the death of the target cell.

Effector d10S cells were activated with a mixture of PMA and ionomycin, an approach known to induce in T killer cells significant levels of cytotoxic activity (Russell 1986; Lancki *et al.* 1987). For d10S cells this activation required MMS for a period of 3 h (this paper), and led to an increased transcription of Fas ligand message (Suda *et al.* 1993). The MMS-requiring activation of d10S cells and the Ca^{2+}-independence of their cytotoxicity (Rouvier *et al.* 1993) are very

reminiscent of recently described characteristics of several CD4$^+$ cytotoxic cell clones (Shih & Bollom 1990; Strack *et al.* 1990; Tite 1990; Ju 1991; Abrams & Russell 1991; Grogg *et al.* 1992; Ozdemirli *et al.* 1992). Among these, some seemed to use a TNF-based mechanism of lysis. Some others behaved in a way suggesting, by comparison with our results, the possibility that they use a Fas-dependent mechanism. For instance, CD4 Th1 clones could be activated by PMA and ionomycin, which required MMS, and then exerted cytotoxicity in a calcium-independent manner (Ozdemirli *et al.* 1992). In addition, in some other systems cytotoxic activity could be induced rather rapidly, as in human blood T cells (Azuma *et al.* 1993).

As discussed before (Rouvier *et al.* 1993), Fas transduction of target cell death implies that the effector cell expresses a functional Fas ligand. MMS may be required, not only for the synthesis of enough of this ligand (Suda *et al.* 1993), but also for the synthesis or the functional availability of any molecule (such as LFA-1, not shown) other than Fas but also required for lysis by d10S cells. Once activated, the d10S cells did not require MMS to induce the death of their target cells; one might speculate that metabolic requirements at the effector cell level may then be minimal. The observation that formaldehyde-fixed activated d10S cells did not exhibit any cytotoxic activity (V. Depraetere, unpublished data) prevents any firm conclusions being drawn.

Cytotoxicity by already activated d10S did not require MMS in the dying target cells, nor was it increased in the presence of MMS inhibitors. Cell death in this case is Fas-transduced (Rouvier *et al.* 1993). Fas-transduced cell death occurring upon engagement of Fas with antibodies has been reported to increase upon inhibition of MMS (Yonehara *et al.* 1989; Itoh *et al.* 1991). Fas-based cell-mediated lysis may not be equivalent to anti-Fas antibody-mediated cell death: cytotoxic T cells may contribute factors other than the Fas ligand. Another, not unlikely possibility is that the nature of the Fas-bearing target cells conditions the effect of MMS inhibitors. These possibilities could be tested by investigating in parallel, on the same Fas-bearing cells, the effect of these antibodies and of activated d10S cells.

As discussed before (Golstein *et al.* 1991), T cell-induced target cell death shares with classical models of apoptosis characteristic morphological features (Sanderson 1976; Don *et al.* 1977; Liepins *et al.* 1977; Matter 1979) and DNA fragmentation (Russell *et al.* 1982; Russell 1983; Duke *et al.* 1983; Cohen *et al.* 1985; Schmid *et al.* 1986; but see Zychlinsky *et al.* 1991). The lack of effect of MMS inhibitors on T cell-mediated cell death is one of the features making this type of apoptosis apparently different from several others. In the present series of experiments this could be demonstrated using the same dying cells: the d10S cell-mediated death of thymocytes did not require the synthesis of any of the molecules whose synthesis was required for the DEX-induced apoptotic death of the same thymocytes. Moreover, although under certain conditions DEX protects cells from antibody-mediated, TCR/CD3-transduced cell death (Zacharchuk *et al.* 1990; Iseki *et al.* 1991; Iwata *et al.* 1991), or from TNF-mediated, TNF-R-transduced cell death (Beyaert *et al.* 1990), it clearly did not protect thymocytes from d10S-mediated, Fas-transduced cell death. It could be that in some cases T cell-mediated 'apoptosis' mechanistically differs, more than just by the initial death signal transduction pathway, from other types of apoptotic death.

We thank M. Nabholz (Lausanne) for the initial gift of PC60 cells and Anne-Marie Schmitt-Verhulst for comments on the manuscript. This work was supported by institutional grants from Institut National de la Santé et de la Recherche Médicale and Centre National de la Recherche Scientifique and by additional grants from the Association pour la Recherche contre le Cancer and the Ligue Nationale Française contre le Cancer.

REFERENCES

Abrams, S.I. & Russell, J.H. 1991 CD4+ T lymphocyte-induced target cell detachment: a model for T cell-mediated lytic and nonlytic inflammatory processes. *J. Immunol.* **146**, 405–413.

Azuma, M., Cayabyab, M., Phillips, J.H. & Lanier, L.L. 1993 Requirements for CD28-dependent T cell-mediated cytotoxicity. *J. Immunol.* **150**, 2091–2101.

Berke, G. 1989 Functions and mechanisms of lysis induced by cytotoxic T lymphocytes and natural killer cells. In *Fundamental immunology* (ed. W. E. Paul), pp. 735–764. New York: Raven Press Ltd.

Beyaert, R., Suffys, P., Van Roy, F. & Fiers, W. 1990 Inhibition by glucocorticoids of tumor necrosis factor-mediated cytotoxicity: evidence against lipocortin involvement. *FEBS Lett.* **262**, 93–96.

Brunner, K.T., Mauel, J., Cerottini, J.-C. & Chapuis, B. 1968 Quantitative assay of the lytic action of immune lymphoid cells on 51Cr-labelled allogeneic target cells in vitro; inhibition by isoantibody and by drugs. *Immunology* **14**, 181–196.

Cohen, J.J. & Duke, R.C. 1984 Glucocorticoid activation of a calcium-dependent endonuclease in thymocyte nuclei leads to cell death. *J. Immunol.* **132**, 38–42.

Cohen, J.J., Duke, R.C., Chervenak, R., Sellins, K.S. & Olson, L.K. 1985 DNA fragmentation in targets of CTL: an example of programmed cell death in the immune system. *Adv. exp. med. Biol.* **184**, 493–508.

Cohen, J.J. 1991 Programmed cell death in the immune system. *Adv. Immunol.* **50**, 55–83.

Conzelmann, A., Corthésy, P., Cianfriglia, M., Silva, A. & Nabholz, M. 1982 Hybrids between rat lymphoma and mouse T cells with inducible cytolytic activity. *Nature, Lond.* **298**, 170–172.

Don, M.M., Ablett, G., Bishop, C.J., Bundesen, P.G., Donald, K.J., Searle, J. & Kerr, J.F.R. 1977 Death of cells by apoptosis following attachment of specifically allergized lymphocytes in vitro. *Austr. J. exp. Biol.* **55**, 407–417.

Duke, R.C., Chervenak, R. & Cohen, J.J. 1983 Endogenous endonuclease-induced DNA fragmentation: an early event in cell-mediated cytolysis. *Proc. natn. Acad. Sci. U.S.A.* **80**, 6361–6365.

Duke, R.C. 1991 Apoptosis in cell-mediated immunity. In *Apoptosis: the molecular basis of cell death* (ed. L. D. Tomei & F. O. Cope), pp. 209–226. New York: Cold Spring Harbor Laboratory Press.

Ellis, R.E., Yuan, J. & Horvitz, H.R. 1991 Mechanisms and functions of cell death. *A. Rev. Cell Biol.* **7**, 663–698.

Golstein, P., Ojcius, D.M. & Young, J.D.-E 1991 Cell death mechanisms and the immune system. *Immunol. Rev.* **121**, 29–65.

Grogg, D., Hahn, S. & Erb, P. 1992 CD4+ T cell-mediated killing of major histocompatibility complex class II-positive antigen-presenting cells (APC). III. CD4+ cytotoxic T cells induce apoptosis of APC. *Eur. J. Immunol.* **22**, 267–272.

Henkart, P.A. 1985 Mechanism of lymphocyte-mediated cytotoxicity. *A. Rev. Immunol.* **3**, 31–58.

Iseki, R., Mukai, M. & Iwata, M. 1991 Regulation of T lymphocyte apoptosis: signals for the antagonism between activation- and glucocorticoid-induced death. *J. Immunol.* **147**, 4286–4292.

Itoh, N., Yonehara, S., Ishii, A., Yonehara, M., Mizushima, S.-I., Sameshima, M., Hase, A., Seto, Y. & Nagata, S. 1991 The polypeptide encoded by the cDNA for human cell surface antigen Fas can mediate apoptosis. *Cell* **66**, 233–243.

Iwata, M., Hanaoka, S. & Sato, K. 1991 Rescue of thymocytes and T cell hybridomas from glucocorticoid-induced apoptosis by stimulation via the T cell receptor/CD3 complex: a possible *in vitro* model for positive selection of the T cell repertoire. *Eur. J. Immunol.* **21**, 643–648.

Ju, S.-T. 1991 Distinct pathways of CD4 and CD8 cells induce rapid target DNA fragmentation. *J. Immunol.* **146**, 812–818.

Kerr, J.F.R., Wyllie, A.H. & Currie, A.R. 1972 Apoptosis: a basic biological phenomenon with wide-ranging implications in tissue kinetics. *Br. J. Cancer* **26**, 239–257.

Kerr, J.F.R. & Harmon, B.V. 1991 Definition and incidence of apoptosis: an historical perspective. In *Apoptosis: the molecular basis of cell death* (ed. L. D. Tomei & F. O. Cope), pp. 5–29. New York: Cold Spring Harbor Laboratory Press.

Lancki, D.W., Weiss, A. & Fitch, F.W. 1987 Requirements for triggering of lysis by cytolytic T lymphocyte clones. *J. Immunol.* **138**, 3646–3653.

Landon, C., Nowicki, M., Sugawara, S. & Dennert, G. 1990 Differential effects of protein synthesis inhibition on CTL and targets in cell-mediated cytotoxicity. *Cell. Immunol.* **128**, 412–426.

Liepins, A., Faanes, R.B., Lifter, J., Choi, Y.S. & De Harven, E. 1977 Ultrastructural changes during T-lymphocyte-mediated cytolysis. *Cell. Immunol.* **28**, 109–124.

Lockshin, R.A. & Zakeri, Z. 1991 Programmed cell death and apoptosis. In *Apoptosis: the molecular basis of cell death* (ed. L. D. Tomei & F. O. Cope), pp. 47–60. New York: Cold Spring Harbor Laboratory Press.

Martin, S.J., Lennon, S.V., Bonham, A.M. & Cotter, T.G. 1990 Induction of apoptosis (programmed cell death) in human leukemic HL-60 cells by inhibition of RNA or protein synthesis. *J. Immunol.* **145**, 1859–1867.

Matter, A. 1979 Microcinematographic and electron microscopic analysis of target cell lysis induced by cytotoxic T lymphocytes. *Immunology* **36**, 179–190.

Ozdemirli, M., El-Khatib, M., Bastiani, L., Akdeniz, H., Kuchroo, V. & Ju, S.-T. 1992 The cytotoxic process of CD4 Th1 clones. *J. Immunol.* **149**, 1889–1895.

Podack, E.R. 1985 Molecular mechanism of lymphocyte-mediated tumor cell lysis. *Immunol. Today* **6**, 21–27.

Podack, E.R., Hengartner, H. & Lichtenheld, M.G. 1991 A central role of perforin in cytolysis? *A. Rev. Immunol.* **9**, 129–157.

Rouvier, E., Luciani, M.-F. & Golstein, P. 1993 Fas involvement in Ca^{2+}-independent T cell-mediated cytotoxicity. *J. exp. Med.* **177**, 195–200.

Ruff, M.R. & Gifford, G.E. 1981 Rabbit tumor necrosis factor: mechanism of action. *Infect. Immun.* **31**, 380–385.

Russell, J.H., Masakovski, V., Rucinsky, T. & Phillips, G. 1982 Mechanisms of immune lysis. III. Characterization of the nature and kinetics of the cytotoxic T lymphocyte-induced nuclear lesion in the target. *J. Immunol.* **128**, 2087–2094.

Russell, J.H. 1983 Internal disintegration model of cytotoxic lymphocyte-induced target damage. *Immunol. Rev.* **72**, 97–118.

Russell, J.H. 1986 Phorbol-ester stimulated lysis of weak and non-specific target cells by cytotoxic T lymphocytes. *J. Immunol.* **136**, 23–27.

Sanderson, C.J. 1976 The mechanism of T cell mediated cytotoxicity. II. Morphological studies of cell death by time-lapse microcinematography. *Proc. R. Soc. Lond.* B **192**, 241–255.

Schmid, D.S., Tite, J.P. & Ruddle, N.H. 1986 DNA fragmentation: manifestation of target cell destruction mediated by cytotoxic T-cell lines, lymphotoxin-secreting helper T-cell clones, and cell-free lymphotoxin-containing supernatant. *Proc. natn. Acad. Sci. U.S.A.* **83**, 1881–1885.

Searle, J., Lawson, T.A., Abbott, P.J., Harmon, B. & Kerr, J.F.R. 1975 An electron-microscope study of the mode of cell death induced by cancer-chemotherapeutic agents in populations of proliferating normal and neoplastic cells. *J. Pathol.* **116**, 129–138.

Shih, C.C.-Y. & Bollom, M. 1990 The acquisition and maintenance of cytolytic activity by CD4$^+$ murine T-lymphocyte clones. *Cell. Immunol.* **130**, 160–175.

Strack, P., Martin, C., Saito, S., Dekruyff, R.H. & Ju, S.-T. 1990 Metabolic inhibitors distinguish cytolytic activity of CD4 and CD8 clones. *Eur. J. Immunol.* **20**, 179–184.

Suda, T., Takahashi, T., Golstein, P. & Nagata, S. 1993 Molecular cloning and expression of the Fas ligand: a novel member of the tumor necrosis factor family. *Cell* **75**, 1169–1178.

Thorn, R.M. & Henney, C.S. 1976 Studies on the mechanism of lymphocyte-mediated cytolysis. VI. A reappraisal of the requirement for protein synthesis during T cell-mediated lysis. *J. Immunol.* **116**, 146–149.

Tite, J.P. 1990 Differential requirement for protein synthesis in cytolysis mediated by class I and class II MHC-restricted cytotoxic T cells. *Immunology* **70**, 440–445.

Trauth, B.C., Klas, C., Peters, A.M.J., Matzku, S., Moller, P., Falk, W., Debatin, K.-M. & Krammer, P.H. 1989 Monoclonal antibody-mediated tumor regression by induction of apoptosis. *Science, Wash.* **245**, 301–305.

Tschopp, J. & Nabholz, M. 1990 Perforin-mediated target cell lysis by cytolytic T lymphocytes. *A. Rev. Immunol.* **8**, 279–302.

Von Boehmer, H., Hengartner, H., Nabholz, M., Lernhardt, W., Schreier, M.H. & Haas, W. 1979 Fine specificity of a continuously growing killer cell clone specific for H-Y antigen. *Eur. J. Immunol.* **9**, 592–597.

Wyllie, A.H. 1980 Glucocorticoid-induced thymocyte apoptosis is associated with endogenous endonuclease activation. *Nature, Lond.* **284**, 555–556.

Wyllie, A.H., Morris, R.G., Smith, A.L. & Dunlop, D. 1984 Chromatin cleavage in apoptosis: association with condensed chromatin morphology and dependence on macromolecular synthesis. *J. Pathol.* **142**, 67–77.

Yonehara, S., Ishii, A. & Yonehara, M. 1989 A cell-killing monoclonal antibody (Anti-Fas) to a cell surface antigen co-downregulated with the receptor of tumor necrosis factor. *J. exp. Med.* **169**, 1747–1756.

Young, L.H.Y., Liu, C.-C., Joag, S., Rafii, S. & Young, J.D.-E 1990 How lymphocytes kill. *A. Rev. Med.* **41**, 45–54.

Zacharchuk, C.M., Mercep, M., Chakraborti, P.K., Simons, S.S.Jr. & Ashwell, J.D. 1990 Programmed T lymphocyte death. Cell activation- and steroid-induced pathways are mutually antagonistic. *J. Immunol.* **145**, 4037–4045.

Zychlinsky, A., Zheng, L.-M., Liu, C.-C. & Young, J.D.-E 1991 Cytotoxic lymphocytes induce both apoptosis and necrosis in target cells. *J. Immunol.* **146**, 393–400.

13

Life, death and genomic change in perturbed cell cycles

ROBERT T. SCHIMKE, ANDREW KUNG, STEVEN S. SHERWOOD, JAMIE SHERIDAN AND RAKESH SHARMA

Department of Biological Sciences, Stanford University, Stanford, California 94305, U.S.A.

SUMMARY

HeLaS3 cells undergo apoptosis after 18–24 h of cell cycle stasis irrespective of the agent employed (colcemid, aphidicolin, cis-platin). At high drug concentrations apoptosis occurs in cells arrested in the cell cycle in which the drug is applied and at a cell cycle position dependent on the mechanism of drug action. At low concentrations (or short exposure times) cells undergo apoptosis after progressing through an aberrant mitosis and only after 18 h of cell cycle stasis in a 'pseudo G1/S' cell cycle position. Aberrant mitoses result in miltipolar mitoses, chromosomal breakage and interchromosomal concatenation events. We propose that the ability of cells to delay progression into aberrant mitosis, as well as their propensity to undergo apoptosis, are important determinants of clinical cytotoxicity. We also suggest that apoptosis plays an important role in preventing the generation of aneuploidy and recombination and rearrangement events commonly associated with cancer.

1. CELL CYCLE STASIS AND APOPTOSIS

We have analysed how various chemotherapeutic agents induce apoptosis, employing agents with different mechanisms of action: (i) colcemid (mitotic spindle blocking agent); (ii) aphidicolin (DNA synthesis inhibitor); and cis-platin (DNA damaging agent) which alter various aspects of cell cycle progression (Sherwood *et al.* 1994*a, b*). We have analysed HeLaS3 cells both in a long-term high dose exposure protocol, typical of cytotoxicity studies with cultured cells, as well as low dose or short term, high dose exposure protocols that are more comparable to those employed clinically, i.e. exposures of the order of 12–24 h duration.

2. HIGH DOSE–CONTINUOUS EXPOSURE

Figure 1 shows flow cytometric analyses of HeLaS3 cells exposed to concentrations of colcemid (figure 1*a*) and aphidicolin (figure 1*b*) that are completely inhibitory (see Sherwood *et al.* 1994*a, b*). In both instances, cells are 'arrested' at a cell cycle position consistent with the mechanism of drug action (early S-phase for aphidicolin, metaphase for colcemid). Cells undergo apoptosis only after a period of cell cycle stasis of 18 h. Apoptosis is discerned in flow cytometry by a reduction in cell size and an apparent decreased in DNA content per cell (arrows). Apoptosis has been documented by microscopic examination of such cells and the demonstration of DNA ladder formation. High dose cis-platin arrests cells in early S-phase from which they undergo apoptosis after an 18 h period of cell cycle stasis (data not shown).

3. LOW DOSE–CONTINUOUS EXPOSURE, OR HIGH DOSE–SHORT TIME EXPOSURE

With drug exposure protocols that more closely mimic clinical treatment régimes, cells undergo apoptosis only after progressing through mitosis. Such mitoses are highly abnormal, and cells remain in a 'pseudo G1/S' position for 18 h prior to commencing apoptosis. Figure 2 shows results with low continuous colcemid (figure 2*a*) or high colcemid for 12 h followed by its removal (figure 2*b*). In both cases (arrows) a population of microcells is generated with less than G1 content. These 'cells' when newly generated contain normal nuclear membranes, decondensed chromatin and exclude vital dyes. They are largely the result of multipolar mitoses as discussed below. Such microcells are unable to progress into a subsequent cell cycle and commence apoptosis 18 h after their generation (Sherwood *et al.* 1994*b*). Low doses of cis-platin likewise allow cells to undergo mitosis, but the resulting G1 cells are unable to progress through a subsequent cell cycle and undergo apoptosis 18 h following mitosis from an early S-phase position (Sherwood *et al.* 1994*b*).

The above results show that the detailed mechanisms of cytotoxicity differ based on the concentration of inhibitory agent employed. At high doses, cell cycle stasis occurs in the same cell cycle in which the agent is applied. Low dose or short exposures of a drug provokes aberrant mitotic events, the consequence of which is cell cycle stasis in G1 or early S-phase. In both instances apoptosis does not occur until cells have been in cell cycle stasis for 18 h.

Figure 3 shows HeLaS3 cells exposed to concentrations of aphidicolin that inhibit progression through

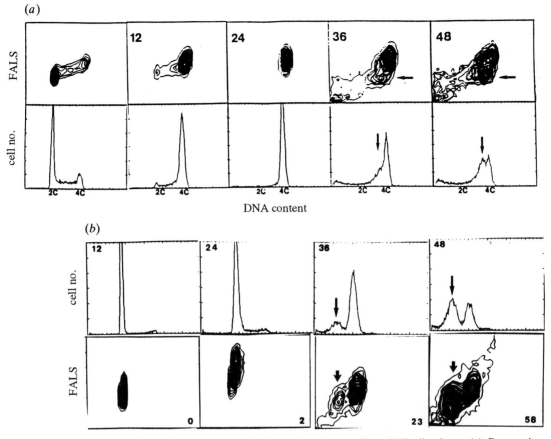

Figure 1. Cell cycle progression and apoptosis at (*a*) high colcemid and (*b*) aphidicolin doses. (*a*) Progression of asynchronous HeLaS3 cells (cell cycle time normally is 18 h) in the presence of 70 ng ml^{-1} colcemid (complete inhibition of mitotic spindle formation and metaphase arrest (Kung *et al.* 1990)). Apoptosis can discerned by a 'reduction' in DNA content per cell and a decrease in cell size-FALS (arrows). DNA is stained with propidium iodide. (*b*) Effects of 5 ng ml^{-1} aphidicolin (complete inhibition of DNA synthesis) in mitotically synchronized cells. Apoptosis is evident at 36 h by a reduction in cell size and DNA content (arrows). The numbers in the lower box indicate the number of apoptotic cells determined microscopically (condensed chromatin).

S-phase to different degrees. At 0.20 ng ml^{-1}, cells progress to a 4C DNA content with an extensive delay (normal HeLaS3 cell cycle time is 18 h) but eventually cells undergo mitosis (60 h) with extensive generation of microcells (small arrow). Although such > 2C 'particles', as detected by flow cytometry, appear similar to those generated at high aphidicolin concentrations (figure 1*b*), microscopic examination indicates high aphidicolin 'particles' are apoptotic cells (chromatin condensation, DNA degradation, permeable membranes) as opposed to microcells. Thus, microscopic examination is necessary to define the nature of > 2C 'particles' which are commonly observed by flow cytometric analyses as cells undergo apoptosis. At progressively higher aphidicolin concentrations (figure 3), the rate of progression through S-phase is proportionally delayed and progressively more cells undergo apoptosis (lower panel). Cells undergoing apoptosis at all concentrations constitute a population derived from early S-phase (large arrows). This finding suggests the existence of a critical early S-phase cell cycle position in which cells are particularly susceptible to undergoing apoptosis. If cells pass beyond this point, they progress, albeit slowly, through the remainder of S-phase without undergoing apoptosis. However, cells that 'escape' the

early S-phase apoptotic position are subject to aberrant mitoses at such time as they undergo mitosis (beyond the time frame of this data) and may undergo apoptosis following an 18 h period of cell cycle stasis as a consequence of aberrant mitotic events (see below).

4. MITOTIC ABERRATIONS AND CELL CYCLE PERTURBATIONS

The consequence of mitotic aberrations is the potential generation of daughter cells with incomplete chromosome complements. Depending on time and concentration parameters, all agents studied (colcemid, aphidicolin, cis-platin) can result in such mitotic aberrations. Any cell, or cell population, may have combinations of such aberrations. Examples are shown in figure 4 and are all derived from cells treated with a low concentration of aphidicolin (0.3 ng ml^{-1}).

(*a*) *Multipolar mitoses*

Figure 4*a,b* shows mitotic HeLas3 cells stained with a beta-tubulin antibody or a centrosome antibody respectively (see arrows). Figure 4*a* shows the formation of multiple spindles, and figure 4*b* shows

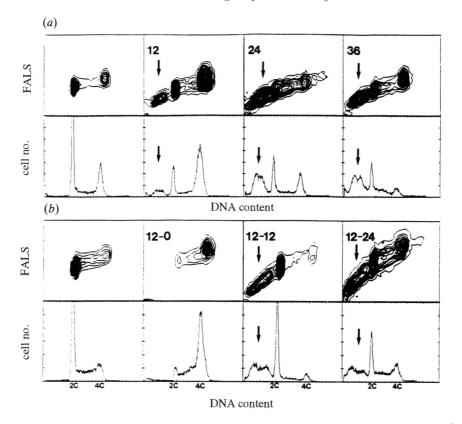

Figure 2. Cell cycle progression of asynchronous HeLaS3 cells in (*a*) continuous presence of 15 ng ml^{-1}, a concentration that does not completely disrupt spindle assembly, and (*b*) exposure to 70 ng ml^{-1} colcemid for 12 h, followed by its removal. Arrows indicate the generation of microcells (see text). Although such 2C 'particles' appear similar in flow cytometry to those generated with APC (figure 1*b*), they are not apoptotic when first generated (see text).

the aberrant metaphase positioning of condensed chromosomes with three centrosomes distributed such that this cell will undergo a tripolar mitosis following karyokinesis and cytokinesis. We suggest that during a slow progression through the cell cycle, centrosome replication/splitting and movement are altered to produce multipolar mitoses (Kreyer *et al.* 1984; Sluder *et al.* 1986).

(*b*) *Chromosomal interconcatenation and aberrant karyokinesis and cytokinesis*

Figure 4*c* shows a typical metaphase from aphidicolin-treated HeLaS3 cells in an attempted metaphase anaphase transition (no mitotic spindle blocking agent was employed). The abnormalities in this metaphase are many, of which we wish to emphasize the inability of chromosomes to separate (both chromatids as well as unrelated chromosomes) which will result in the inability of chromosomes to undergo proper segregation. We interpret this type of finding to be the result of interchromatid and interchromosome concatenation. Although the existence of interchromatid concentration has been recognized to occur during mitosis of cells treated with topoisomerase II inhibitors (Downs *et al.* 1991), we observe such a phenomenon with virtually any agent that inhibits cell cycle progression. In addition to interchromatid concatenation, concatenation also

occurs between unrelated chromosomes. Interchromosomal concatenation has not been emphasized previously, although its existence has been reported (Schmid *et al.* 1983; Rose & Holm 1993). Our observations lead us to conclude that interchromosomal concatenation is a common result of perturbation of cell cycle progression that contributes to eventual cell death. This metaphase (figure 4*c*) also shows chromosomal breakage and the suggestion of attempted multipolar segregation. Thus, various aberrations of chromosome and chromatid integrity and spindle formation can occur in the same metaphase. Figure 4*d* shows a typical consequence of chromosome concatenation in G1 daughter cells following mitosis, where nuclear morphology is abnormal, cells contain micronuclei, and where daughter cells cannot separate completely because of concatenated DNA stretching (arrow) across of the cytokinetic furrow. Extensive concatenation can result in the inability of cells to undergo karyokinesis and cytokinesis during an attempted mitosis, or a variable karyokinesis such that most of the chromosomes are segregated into a single daughter cell. Such cells appear in flow cytometry as 4C cells which can progress into further cell cycles. Figure 4*e* shows such a cell which contains a large, abnormally lobulated nucleus. Note also that this cell contains micronuclei indicative of either chromosome loss or chromosome fragment loss from the nucleus following nuclear membrane reassembly.

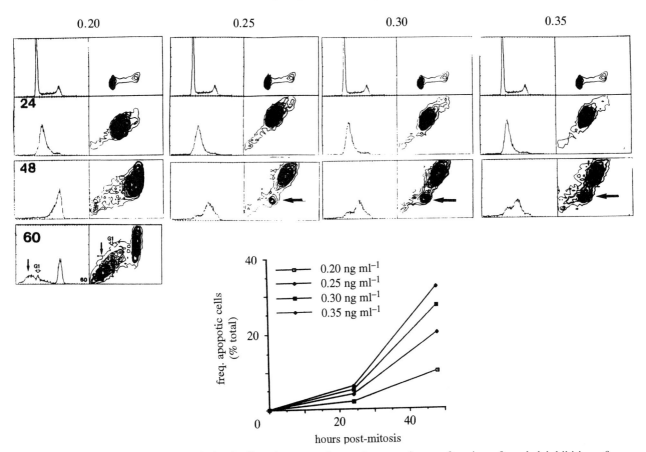

Figure 3. Flow cytometric analysis of cell cycle progression and apoptosis as a function of graded inhibition of DNA synthesis by varied aphidicolin concentrations. Asynchronous HeLaS3 cells were treated continuously with varying concentrations $(ng\,ml^{-1})$ of aphidicolin. The large arrows indicate apoptotic cells within the cell populations. The small arrow (at low APC concentration) constitutes predominantly microcells following a delayed mitoses. The lower panel shows quantification of apoptotic cells (chromatin condensation: apoptotic bodies visualized by light microscopy). Cell sorting confirmed that apoptotic cells are those indicated by the large arrows.

Such cells are not subject to apoptosis following mitosis (as opposed to those which go through karyokinesis and cytokinesis). We attribute this difference to the fact that such tetraploid or aneuploid cells contain a genome complement that can sustain further cell proliferation, whereas aberrant G1 and microcells do not have complete genome complements. We suggest that this process may be an important means of generating aneuploidy as it occurs in cancers.

(c) *Chromosome breakage*

Variable degrees of chromosome breakage or presence of extrachromosomal DNA are also characteristic findings of cells treated with various agents that result in slow progression towards and into mitosis. The metaphase of figure 4c shows chromosome breakages (see also Schimke *et al.* 1988; Sherwood *et al.* 1988). Although such breaks are typically attributed to the direct action of an agent, we suggest that chromosome breaks can also be accounted for as the result of aberrant progression into mitosis *per se* prior to completion of chromosome condensation and chromosome deconcatenation (see Kung *et al.* 1993).

5. THE IMPORTANCE OF THE MITOTIC CHECKPOINT IN CELL SURVIVAL AND IN GENOME STABILITY

Our results emphasize the role of mitosis in cell death and genomic stability, in particular when cells are subject to discontinuous and low exposure régimes. We suggest that transient exposures to carcinogenic agents may do likewise. Thus, the potential ability of cells to prevent aberrant mitoses may be important in clinical chemotherapy resistance and in maintaining genome stability (Schimke *et al.* 1991; Hartwell 1993). The mitotic checkpoint, as defined in yeast, involves the 'sensing' of completion of DNA synthesis, chromosome and mitotic spindle integrity to delay or prevent aberrant or lethal mitoses. In the absence of such 'sensors', yeast are reproductively non-viable following aberrant mitotic events (Weinert & Hartwell 1987; Murray 1992).

Checkpoint controls analogous to those of yeast exist in mammalian cells. Whereas a number of human cell lines, including HeLaS3, have mitotic checkpoint properties analogous to those of yeast, rodent cell lines often lack such mitotic checkpoint control properties (Kung *et al.* 1990, 1993; Schimke 1991). Perhaps most striking is cytotoxicity of CHO,

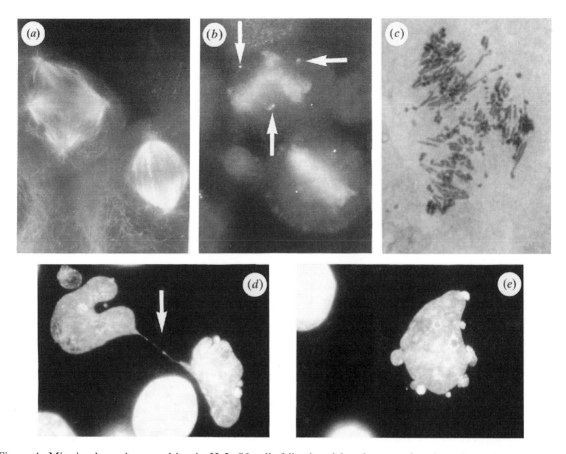

Figure 4. Mitotic aberrations resulting in HeLaS3 cells following delayed progression through a cell cycle. Cells were treated with 0.25 μg ml⁻¹ aphidicolin. (*a*) Mitotic cells stained with anti-tubulin, showing multiple centres of mitotic spindle nucleation in one of two cells. (*b*) Mitotic cell stained with an anti-centrosome antibody. The rhodamine stain shows aberrant chromosome positioning at the metaphase plate, as well as centrosomes placed abnormally (arrows). (*c*) A metaphase–anaphase transition (giemsa). This spread shows a composite of a number of abnormalities including: (i) inability of chromosomes to separate properly; (ii) lack of attachment of chromosomes to mitotic spindles; and (iii) extensive chromosomal breakage. (*d*) Daughter cells which have undergone incomplete karyokinesis and cytokinesis and which are attached through DNA strands (arrow) (Hoechsts 33342). The G1 nuclei are abnormal. Note also several micronuclei of varying sizes. (*e*) A HeLaS3 cell that has progressed through mitosis, but has failed to successfully undergo karyokinesis or cytokinesis (Hoechsts 33342). Note in this cell the presence of micronuclei, as well as a large and lobulated nucleus.

but not HeLaS3 cells, to an 18 h period of complete inhibition of DNA synthesis by aphidicolin. Death in CHO cells is a consequence of aberrant mitoses following removal of aphidicolin and the difference between HeLaS3 and CHO cells is attributed to the a down-regulation of overall protein synthesis in HeLaS3, but not CHO cells, during a period of inhibition of DNA synthesis. CHO cells accumulate cyclin B to mitotically competent levels during S-phase inhibition whereas HeLaS3 cells do not. As a consequence, CHO cells undergo aberrant (early) mitoses whereas HeLaS3 do not (Kung *et al.* 1993). Thus, the control of progression into an aberrant mitosis constitutes the lethal event. One consequence of aberrant mitoses that result from an 18 h exposure to high aphidicolin in CHO cells is the generation of extensive chromosomal breakage and an increased frequency of gene amplification in the clonogenic survivors of such a treatment (Sherwood *et al.* 1988; Schimke *et al.* 1988).

Intactness of the mitotic checkpoint is important in

two contexts related to cancer biology based on our studies with cultured cells.

(*a*) Cytotoxicity in relation to resistance

When (if) cancers are exposed to short-term or minimally effective drug concentrations, aberrant mitoses are a central feature of cytotoxicity. Therefore the capacity of cells to 'withstand' entry into potentially lethal mitoses within the dose–time frame of clinical drug administration constitutes a mechanism(s) of resistance. We wish to point out that among such parameters of mitotic regulation are: (i) down regulation of cyclin B accumulation (Kung *et al.* 1993); (ii) intactness of a G2 (radiation) checkpoint (O'Conner & Kohn 1992); (iii) ability to deconcatenate DNA; and (iv) lack of centrosome replication or splitting during delayed cell cycle progression. These properties vary among cell lines we have studied (only HeLaS3 data have been presented herein), and likely vary in cancers as well. Thus, resistance mechanisms as they occur in cancers may be determined, in part, by how cells respond to an inhibitory agent as opposed to current concepts of resistance mechanisms that equate resistance with mechanisms that prevent drug action. We note that the latter mechanisms have generally been elaborated in experimental cell culture systems subjected to high, continuous drug exposure protocols.

(b) Genome instability

We propose that when cells are slowed down in progression through a cell cycle, among the consequences is chromosome breakage and altered karyokinesis. Altered karyokinesis can result in aneuploidy, a central theme of many cancers. In addition, chromosomal breakage is proposed to be an initial event in recombinational processes resulting in gene amplification in many models (Schimke *et al.* 1988; Windle & Wahl 1992; Chi *et al.* 1993), another common theme in cancers. We have recently reported a correlation between the propensity for gene amplification in different cultured mammalian cell lines and their mitotic checkpoint properties (Sharma & Schimke 1994).

6. APOPTOSIS: THE ULTIMATE 'REPAIR' FOR PERTURBED CELL CYCLES

Within the context of cancer biology, all apoptotic deaths are desired (Kerr *et al.* 1972). We suggest that, in addition to mitotic checkpoint properties (see above), the inherent ability of cells to undergo apoptosis within clinical chemotherapy exposure times may be an additional, critical determinant in successful chemotherapy (Vaux 1993). We have found that cell lines commence apoptosis subsequent to metaphase arrest (colcemid) at times varying from 6 h to 30 h (L. Kim & L. Yilmaz, unpublished results). Interestingly, cell lines that undergo apoptosis most rapidly are predominately derived from bone marrow precursors (B and T cells) and embryonal cells. This finding suggests that success in cancer treatment may, in part, be a function of ability to initiate apoptosis within the constrained time of patient exposure to drugs and may be dependent on the expression state of genes affecting apoptosis, including Bcl-2 (Sentman *et al.* 1991) and p53 (Yonish-Rouach *et al.* 1991; Lowe *et al.* 1993).

In addition, we suggest that the ease of initiating apoptosis during cell cycle perturbation may also play a role in preventing genomic instability. In our studies with HeLaS3 cells, the treatment conditions most conducive to the generation of aneuploidy and chromosome breakage involve a slowing down of cell cycle progression. Although our studies have concentrated on use of aphidicolin, virtually all agents that interact with DNA (i.e. mutagenic agents) also slow down DNA synthesis and result in aberrant mitotic events. We note (see figure 3) an early S-phase cell cycle position from which cells undergo apoptosis readily. If cells progress past this cell cycle position, they can ultimately progress into aberrant mitoses and

potentially generate aneuploidy and/or recombinational chromosome repair, including deletions, gene amplifications and translocations. The vast majority of such breakage and recombination events will be lethal or neutral. However, an occasional event may contribute to cancer progression. Thus, we suggest that the ability of cells to undergo apoptosis prior to entry into aberrant mitoses when cell cycle progression is altered by any variety of agents constitutes an important means of maintaining genome stability. By virtue of removing potentially 'dangerous' cells, apoptosis constitutes the ultimate 'repair' process for multicellular organisms where loss of any single proliferating cell is inconsequential to the organism. Whether p53 plays such a role in its facilitation of apoptosis under our conditions of cell cycle perturbation is under current study. It is intriguing to suggest that p53 may play a role in facilitating apoptosis at a G1/S cell cycle position where Kastan *et al.* (1991) have shown a role of p53, a cell cycle position which our data with aphidicolin suggests to be a critical position for apoptosis when cells are progressing slowly through a cell cycle.

REFERENCES

Chi, M.S.M., Trask, B. & Hamlin, J.L. 1993 Sister chromatid fusion initiates amplification of the DHFR gene in Chinese cells. *Genes Dev.* **7**, 605–615.

Downs, C.S., Mulinger, A.M. & Johnson, R.T. 1991 Inhibitors of DNA topoisomerase II prevent chromatid separation in mammalian cells but do not prevent exit from mitosis. *Proc. natn. Acad. Sci.* **88**, 8895–8899.

Hartwell, L.H. 1992 Role of yeast in cancer research. *Cancer* **69**, 2615–2619.

Kastan, M.B., Onyekwere, O., Sidransky, D., Vogelstein, B. & Craig, R.W. 1991 Participation of p53 protein in the cellular response to DNA damage. *Cancer Res.* **51**, 6304–6310.

Kerr, J.F.R., Wyllie, A.H. & Currie, A.R. 1972 Apoptosis: a basic biological phenomenon with wide ranging implications in tissue kinetics. *Br. J. Cancer* **26**, 239–257.

Kreyer, G.H.R. & Borisy, H.H. 1984 Centriole distribution during tripolar mitosis in Chinese hamster ovary cells. *J. Cell Biol.* **98**, 2222–2229.

Kung, A.L., Sherwood, S.W. & Schimke, R.T. 1990 Cell line-specific differences in the control of cell cycle progression in the absence of mitosis. *Proc. natn. Acad. Sci. U.S.A.* **87**, 9553–9557.

Kung, A.L., Sherwood, S.W. & Schimke, R.T. 1993 Differences in the regulation of protein synthesis, cyclin b accumulation, and cellular growth in response to the inhibition of DNA synthesis in Chinese hamster ovary and HeLaS3 cells. *J. biol. Chem.* **268**, 23072–23080.

Lowe, S.W., Schmitt, E.M., Smith, S.W., Osborne, B.A. & Jacks, T. 1993 p53 is required for radiation-induced apoptosis in mouse thymocytes. *Nature, Lond.* **362**, 847–849.

Murray, A.W. 1991 Creative blocks: cell cycle checkpoints and feedback controls. *Nature, Lond.* **359**, 599–604.

O'Conner, P.M. & Kohn, K.W. 1992 A fundamental role for cell cycle regulation in the chemosensitivity of cancer cells? *Semin. Cancer Biol.* **3**, 409–416.

Rose, D. & Holm, C. 1993 Meiosis-specific arrest revealed in DNA topoisomerase II mutants. *Molec. Cell. Biol.* **13**, 3445–3455.

Schimke, R.T., Hoy, C., Rice, G., Sherwood, S.W. & Schumacher, R.I. 1988 Enhancement of gene amplification by perturbation of DNA synthesis in cultured mammalian cells. *Cancer Cells* **6**, 317–323.

Schimke, R.T., Kung, A.L., Rush, D.F. & Sherwood, S.W. 1991 Differences in mitotic control among mammalian cells. *Cold Spring Harb. Symp. Quant. Biol.* **56**, 417–425.

Schmid, M., Grunert, D., Haaf, T. & Engle, W. 1983 A direct demonstration of somatically paired heterochromatin of human chromosomes. *Cytogenet. Cell. Genet.* **36**, 554–561.

Sentman, C.L., Shutter, J.R., Hockenbery, D., Kanagawa, O. & Korsmeyer, S.J. 1991 Bcl-2 inhibits multiple forms of apoptosis but not negative selection in thymocytes. *Cell* **67**, 879–885.

Sharma, R.C. & Schimke, R.T. 1994 The propensity for gene amplification: a comparison of protocols, cell lines, and selection agents. *Mutation Res.* **304**, 243–260.

Sherwood, S.W., Schumacher, S.I. & Schimke, R.T. 1987 Effect of cycloheximide on development of methotrexate resistance in Chinese hamster ovary cells treated with inhibitors of DNA synthesis. *Molec. Cell. Biol.* **8**, 2822–2827.

Sherwood, S.W. & Schimke, R.T. 1994 Induction of apoptosis by cell-cycle specific drugs. In *Apoptosis:* *Proceedings of the 5th Pezcoller Foundation Symposium* (ed. H. Mihich & R. T. Schimke), pp. 223–236. New York: Plenum Press.

Sherwood, S.W., Sheridan, J. & Schimke, R.T. 1994 Cell cycle correlates of drug-induced apoptosis in HeLa cells. (Submitted.)

Sluder, G., Miller, F.J. & Rieder, C.L. 1986 The reproduction of centrosomes: nuclear versus cytoplasmic controls. *J. Cell Biol.* **103**, 1873–1881.

Vaux, D. 1993 Toward and understanding of the molecular mechanisms of physiological cell death. *Proc. natn. Acad. Sci. U.S.A.* **90**, 786–789.

Windle, B.E. & Wahl, G.M. 1992 Molecular dissection of mammalian gene amplification: new mechanistic insights revealed by analysis of very early events. *Mutation Res.* **276**, 199–205.

Weinert, T.A. & Hartwell, L.H. 1988 The RAD 9 gene controls the cell cycle response to DNA damage in *Saccharomyces cerevisiae*. *Science, Wash.* **241**, 317–322.

Yonish-Rouach, E., Resnitzky, D., Lotem, J., Sachs, L., Kimchi, A. & Oren, M. 1991 Wild-type 53 induces apoptosis of myeloid leukaemic cells that is inhibited by interleukin 6. *Nature, Lond.* **352**, 345–348.

14

Apoptosis and cancer chemotherapy

J. A. HICKMAN[1], C. S. POTTEN[2], A. J. MERRITT[1,2] AND T. C. FISHER[1]

[1]*Cancer Research Campaign Molecular and Cellular Pharmacology Group, The School of Biological Sciences, University of Manchester, Stopford Building (G38), Manchester M13 9PT, U.K.*
[2]*CRC Department of Epithelial Biology, Paterson Institute for Cancer Research, Christie Hospital (NHS) Trust, Wilmslow Road, Manchester M20 9BX, U.K.*

SUMMARY

The major disseminated cancers remain stubbornly resistant to systemic therapy. Drug-resistant tumours include both slow and fast growing types, with the carcinomas constituting the major problem. Strategies for drug discovery have, in the past, been focused on attempts to design antiproliferative agents, largely targeted to interfere with DNA integrity and replication. The malignant phenotype might be characterized by the emergence of cell populations with a greater survival potential: a lower proclivity to undergo apoptosis. This idea provides a possible explanation of the genesis and progression of cancer and of the inherent resistance of tumour cells to engage apoptosis. Work is described which identifies the molecular basis for differences in the survival potential of stem cells in the crypts of the colon and small intestine. The advantageous survival of colonic stem cells, provided by expression of bcl-2 and a muted p53 response to DNA damage, allows damaged cells to survive. Continued expression of bcl-2 renders tumour cells resistant to drug-induced DNA damage by a mechanism different from classical mechanisms of drug resistance. The attenuation of cell survival is described as a key component in strategies for the drug treatment of disseminated cancers.

1. INTRODUCTION

Cancer pharmacologists continue to struggle with the quest to discover cellular targets at which innovative drug molecules might be aimed. The past four decades have seen only modest success in this endeavour, with the major human cancers continuing to reap significant numbers of lives. The vast increase in knowledge regarding the molecular basis of malignancy, facilitated by advances in molecular biology, has provided a plethora of new potential targets for the aspiring drug discoverer. However, few of these afford an immediately obvious and selective means of preventing tumour growth without affecting that of normal tissues. Moreover, the work of Foulds (1958) has long alerted us to the fact that not one but a series of genetic changes are likely to have occurred prior to the genesis of a frankly malignant cell; on which one of these multiple events should the drug discoverer focus attention? Is there a hierarchy among these events, so that a single 'magic bullet' could be aimed at a critical biochemical feature underpinning the malignant phenotype? And if that critical target were to be modulated by a drug, what should we expect as the end-point of therapy: restoration of normal cellular function, cytostasis of the malignant cell, or its death? It is in asking this latter question that a profound change in perception of the problems of chemotherapy has recently taken place, a change consequent on our understanding ·that targeting a drug to a key feature of cellular biochemistry is only a first step, with subsequent steps and the final outcome being determined by a cellular response to a drug-induced change in homeostasis (Dive & Hickman 1991). Observations that many types of current anticancer drugs, with completely disparate cellular targets, induced apoptosis in susceptible cells (reviewed by Hickman 1992; Dive & Wyllie 1993) played a key feature in this thinking, and suggested that disruption of homeostasis and/or cellular damage initiated a cellular suicide response, or not, according to phenotype. Thus, the end-point of drug therapy was dependent on a cellular response and not solely on the quality or quantity of change induced by the drug. Here lay a possible explanation of what has been termed 'inherent' drug resistance: those tumours that have responded well to therapy may be derived from cells whose response to damage was to engage apoptosis readily; those which were resistant either do not receive the stimulus for cell death (classical mechanisms of drug resistance) or, more importantly, do not respond to it. Our failure to affect many of the solid tumours may reflect the high survival potential of their cells of origin in comparison to tissues like the bone marrow. Evidence to support this idea is described here.

The implications stemming from the idea that the end point of the action of drugs may be dependent upon a cellular response (simple stimulus–response coupling, figure 1) are considerable. First, as stated

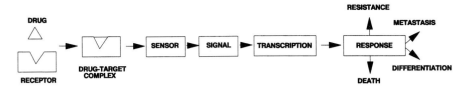

Figure 1. Stimulus response coupling. Drug-induced perturbations of cellular metabolism are presumably 'sensed' by the cell and signals initiated to engage the appropriate response, according to phenotype. One of these responses is to engage apoptosis. It is suggested that at the 'coupling' stage there is modulation of the signal. The outcome of drug treatment is therefore dependent upon the nature of these modulatory events.

above, the failure of drugs to impact on many of the major tumours may be a reflection of the failure of the tumour cells to undergo apoptosis at a threshold of cellular perturbation below that triggering the death of cells in normal tissue, such as the bone marrow or the epithelia of the small intestine. Essentially, this idea suggests that these refractive tumours are resistant to the process of engaging cell death, not to the drugs per se. Secondly, the conundrum of choosing a target from among the many progressive changes in cellular biochemistry characterizing the progression to malignancy may be resolved: selective perturbation of *any* of these changes may be imposed as long as the cell is able to couple this perturbation to the engagement of apoptosis. Is this realizable?

Critical to making an advance is the identification of those factors which modulate the cellular response to damage: the cellular threshold defining whether to survive or to die. The goal of inhibition of cellular survival has the added advantage that it is not necessarily dependent on cell proliferation biochemistry, an advantage important to the therapy of the many slow growing tumours with a low growth fraction.

We, like others, have tested the idea that modulation of the coupling of the response to damage may play a role in drug resistance (Fisher *et al.* 1993; Merritt *et al.* 1994a). The therapy of colon carcinoma was of particular interest for two reasons. First, because it is a tumour for which chemotherapy is only palliative. Secondly, because a study of the tissue from which it originates provided insight to the role that modulation of apoptosis may provide in carcinogenesis and the survival of damaged cells which appear to be able to withstand the further damage imposed by chemotherapeutic drugs and radiation.

2. COLON CARCINOGENESIS: INSUFFICIENT ALTRUISTIC APOPTOSIS

U.K. statistics, from the Cancer Research Campaign, estimated that 28 590 cancers arose in the colon in 1987, accounting for a tenth of deaths from cancer in 1991. A puzzle of bowel cancer has been the observation that less than 5% of cancers arise in the small intestine, with the majority in the colon and rectum (Goligher 1980). Elegant studies comparing the kinetics of cell proliferation in small intestine and colon, both in the rodent and in man, have not

revealed differences which may account for the greater incidence of colon cancer (reviewed by Potten 1992). Rather, morphological studies have revealed that patterns of cell loss by apoptosis from the small intestine and colon are quite distinctive and provide a possible explanation of the differential cancer incidence (Potten *et al.* 1992). Our investigations suggest that the survival of damaged stem cells in the colon may allow the expansion of a population with a greater survival potential than the small intestine, and that maintenance and/or an increase of this survival potential may render the tumour inherently chemoresistant.

(a) Stem cell apoptosis is restricted to the small intestine

Careful analysis of cell positional behaviour and hierarchies in the crypts of the murine colon and small intestine by Potten and colleagues (reviewed in Potten 1992) suggested that the topology of cell renewal in colonic and small intestinal crypts was quite different (figure 2). In the small intestine, the putative stem cells were assignable to cell positions 4 or 5 (figure 2) whereas in the colon, stem cells were positioned at the very base of the colonic crypt, from positions 1 to 2. As in all dynamic tissues, homeostasis in the intestinal crypts appears to be maintained by removing cells excess to requirement by apoptosis. Most interestingly, the pattern of this spontaneous apoptosis was also different in the small intestine and colonic crypts: in the small intestine, apoptotic cells were observed to be restricted to the stem cell region whereas in colonic crypts spontaneous apoptosis was less frequent and occurred in a less topologically restricted way (reviewed by Potten 1992). Critically, few apoptotic cells were observed at the base of colonic crypts, harbouring the stem cells. This differential pattern of apoptosis was exaggerated when animals were exposed to DNA damaging agents or radiation (Ijiri & Potten 1987).

These observations of the differential amounts and position of apoptosis in the crypts led to the hypothesis that damaged small intestinal stem cells were deleted by an 'altruistic' apoptosis whereas in the colon, damaged cells survive and may go on to give rise to cancers (Potten *et al.* 1992). Perhaps the most critical step in carcinogenesis is that a cell capable of replication is able to survive a transforming mutation and to progress through further genetic changes because of a continued, enhanced ability to sustain

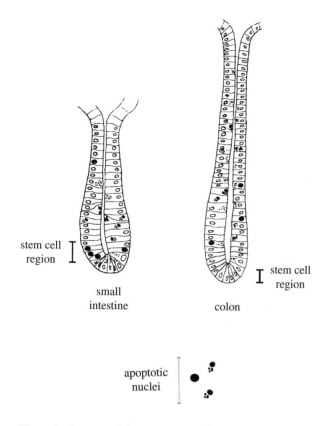

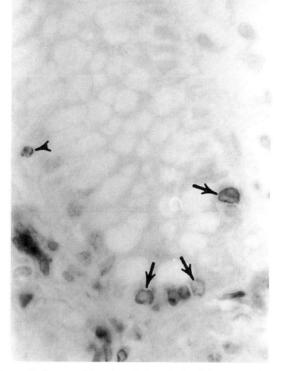

Figure 3. Immunohistochemical analysis of the expression of the human bcl-2 protein, by immunoperoxidase staining, in a normal colonic crypt.

Figure 2. Cartoon of the appearance of longitudinal sections of the crypts from the murine small intestine and colon. The stem cell regions are shown together with the positions of cells which are observed to undergo spontaneous or induced apoptosis. The stem cell in the small intestine is considered to be at position 4 up from the bottom, above the Paneth cells, and at position 1 in the colon, at the very base of the crypt.

and survive damage. This damage may include that imposed by radiation therapy or chemotherapeutic drugs. What is the genetic basis of this survival advantage and how might it relate to chemoresistance?

(b) Bcl-2 is differentially expressed between the small intestine and colon

The idea that the expression of certain genes could modulate survival was made a reality by the discovery that the activity of the gene bcl-2 was to inhibit apoptosis (reviewed by Reed 1994). Using established methods to prepare longitudinal sections of mouse and human colonic and small intestinal crypts, the expression of bcl-2 was investigated by us using immunohistochemistry (Merritt et al., unpublished data). We found that expression of bcl-2 was confined to the colonic epithelia, with little staining in the small intestine. Peroxidase staining of sections from normal human colonic crypts suggested strong perinuclear staining in epithelial cells at the base of the crypt, precisely among the stem cell population (figure 3). The sporadic staining of the epithelial cells of the small intestine was not confined to the stem cell region, and may have been of intraepithelial

lymphocytes. Thus bcl-2 expression was in those cells which resist either spontaneous or induced apoptosis. The result contrasted with a pattern of staining reported by Hockenbery et al. (1991) who claimed that the gene was expressed equally in the crypts of the small intestine and colon. Their data are difficult to reconcile with observations of topographically restricted patterns of both spontaneous and induced apoptosis (see figure 2). The selective expression of bcl-2 presumably protects epithelial stem cells from the colon from excessive loss induced by concentrating toxins from the diet or from the colonic flora.

(c) Colon adenocarcinomas express bcl-2

In a limited but on-going survey of both human and murine colonic tumours, we found significant expression of bcl-2. Interestingly, immunofluorescence staining showed bcl-2 protein to be largely perinuclear in distribution, with occasional intranuclear expression (Merritt et al., unpublished data). Very significant staining of bcl-2 protein has been reported recently in other solid tumours: in carcinomas of the breast (Leek et al. 1994), small cell lung cancer (Pezzella et al. 1993; Ikegaki et al. 1994), androgen-independent prostate cancer (McDonnell et al. 1992; Colombel et al. 1993) and in neuroblastomas (Castle et al. 1994). In the studies of breast carcinomas (Leek et al. 1994), it appeared that the loss of bcl-2 expression in the tumours with a poorer prognosis was associated with the emergence of positive immunohistochemical staining for p53, c-erbB-2 and EGFR (epidermal growth factor receptor). This may be interpreted as representing progression to a higher survival threshold than that provided by bcl-2 alone, and that in this process of tumour progression, the expression of bcl-2, initially required for enhanced survival, may ultimately be lost. Interestingly, in the breast carcinomas there was an inverse relationship between the expression of p53 and bcl-2 (Leek *et al.* 1994). As described below, this also appears to hold for the crypt epithelia of the gut.

(d) *Expression of* bcl-2 *prevents cell death induced by drugs used to treat colon cancer*

The model of stimulus–response coupling discussed above (figure 1) predicts that the expression of genes which modulate apoptosis, like bcl-2, should modulate the response of cells to drug-induced damage. As a mechanism of drug resistance, this is novel. Previous paradigms for drug resistance have included only changes to what is here called the 'stimulus': the amount and quality of damage delivered. The cell has been perceived to be active only in modulating the amount of this stimulus, either by inactivating the drug or effluxing it, by changing the 'target' (for example by utilizing alternative biochemical pathways, or by quantitative or qualitative changes in the target) or by repairing drug-induced damage, for example to DNA.

We chose to study whether the ectopic expression of bcl-2 could provide resistance to a drug class without changing the 'stimulus' in any way. It was therefore essential that we could define and quantitate the stimulus. For this reason we investigated inhibitors of the enzyme thymidylate synthase. There are very specific inhibitors of this enzyme, belonging to different chemical classes of agent with different transport routes across the cell membrane. In addition, the amount of enzyme and its activity can readily be measured, and the results of its inhibition can be quantitated (by changes in pools of thymidine and other nucleotides). After treatment of cells with inhibitors of thymidylate synthase, reduction of pools of thymidine result in a gradual incorporation of deoxyuridine into nascent DNA. This leads to an accumulation of strand breaks, which again can be quantified. Indeed, it has generally been supposed that these breaks are the lethal event induced by this class of drug: DNA breaks in some critical, but as yet unidentified gene(s) proving to be lethal to the cell. With respect to our studies of cell death and carcinogenesis in the colon, these drugs, and particularly 5-flourouracil, are the major class of chemotherapeutic agents administered to patients with colon cancer and we were keen to discover what effect the expression of *bcl*-2 might have on drug sensitivity.

As an easily manipulated model system we used human lymphoma cells transfected with human *bcl*-2 or a vector control (Fisher *et al.* 1993). The transfected cells were found to contain about three times the amount of bcl-2α protein found, by Western blotting, in a naturally *bcl*-2 expressing lymphoma. Both the transfects divided at the same rate and had an equal cell cycle time (24 h) and distribution, important when comparing S-phase-specific drugs. In a detailed study of the effects of 5-fluorodeoxyuridine, it was shown that transfection of *bcl*-2 did not alter the

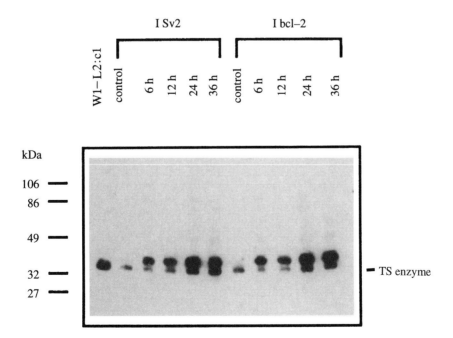

Figure 4. Western blot of the thmidylate synthase enzyme in human B-cell lymphoma cells transfected with either vector (I Sv2) or human bcl-2 (I bcl-2). The blot shows changes in the amount of enzyme after various times of incubation of the cells with 5-fluorodeoxyuridine, as the cells attempt to overcome the blockade of the enzyme. Also, with time, the blot shows the appearance of the higher molecular mass form of thymidylate synthase formed as the drug forms a ternary complex. (The lane WI-L2:c1 contains purified thymidylate synthase protein.) (From Fisher *et al.* (1993), used with permission.)

amount of drug binding, via the formation of a ternary complex to the thymidylate synthase enzyme, as detected by Western blotting, nor was the inhibition-stimulated up-regulation of the enzyme inhibited (figure 4). Analysis, by alkaline elution, of DNA damage in nascent DNA showed that, with time, strand breaks occurred in both the vector and bcl-2 transfected cells (figure 5). Twelve hours after removal of 1 μM of the drug the cells had equivalent amounts of DNA damage, and this continued to increase, with the bcl-2 transfectants apparently accumulating more damage (figure 5). This amount of DNA damage would normally be expected to be lethal to both vector transfects and bcl-2 transfects. In fact, whereas cells transfected with the vector alone readily underwent apoptosis after a 24 h treatment with 5-fluorodeoxyuridine, the bcl-2 expressing cells remained viable (figure 5). Similar results were observed using a series of quinazoline-based thymidylate synthase inhibitors which differed with respect to their membrane transport (passive or facilitated) and their intracellular metabolism: these compounds either were or were not capable of undergoing polyglutamation. Overall, our data clearly showed that *bcl-2* expression prevented cell death independently of the classical mechanisms of drug resistance associated with this class of agent.

The survival of cells with accumulating DNA damage (figure 5) is remarkable. It remains to establish the fate of these cells and what the integrity of their DNA might be at later times. Since drug-treated bcl-2 transfectants repopulated the cultures

(Fisher *et al.* 1993) we presume that DNA was repaired. Whether this repair was with high fidelity, or not, is a very important question: inhibition of thymidylate synthase has previously been shown to increase homologous recombination activity, and it could be presumed that such an event, in cells surviving with high levels of DNA damage, might promote the development of further mechanisms of resistance.

Bcl-2 has now been reported to inhibit cell death induced by a whole variety of chemotherapeutic drugs with different loci of action. This includes glucocorticoids, methotrexate, etoposide (Miyashita & Reed 1993), nitrogen mustard and camptothecin (Walton *et al.* 1993) and X-irradiation (Collins *et al.* 1992). This truly pleotropic resistance extends far beyond the range of drugs to which the so-called multidrug resistance phenotype applies (reviewed by Twenty-man 1992). Importantly, in etoposide-treated CH31 murine B cells transfected with *bcl-2* there was shown to be an increase in the clonogenic potential of the resistant *bcl-2* containing cells, although like our results with fluorodeoxyuridine (above) no change was observed in the amount of DNA damage induced nor, critically, in the rate of repair (Kamesaki *et al.* 1993).

These experiments support the idea that the sensitivity of tumour cells to drugs is not solely dependent on the type or quantity of perturbation delivered by the drugs: the consequences of this for the designer of new drugs are considerable, as outlined below.

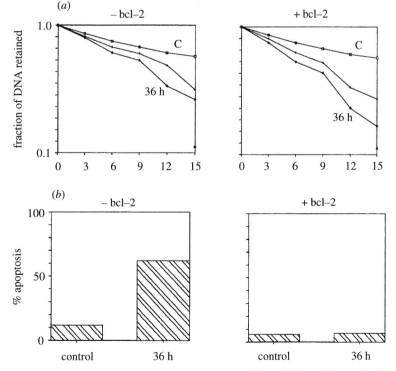

Figure 5. Composite showing (*a*) DNA strand breaks, measured by alkaline elution, in the nascent DNA of 5-fluorodeoxyuridine treated (1 μM for 24 h) human B lymphoma cells which had been transfected with either vector alone (left hand panels) or human bcl-2 (right hand panels) (C = control). (*b*) The percentage of apoptotic cells, measured by acridine orange fluorescence, 36 h after a 24 h treatment with 1 μM fluorodeoxyuridine. Modified from Fisher *et al.* (1993), used with permission.

(e) p53 is differentially expressed in the crypts of the small intestine and colon: further implications for chemotherapy of colonic tumours

In defining the accumulation of genetic changes that characterize the progression of colon cancer, the loss of function of p53, the so-called 'guardian of the genome' (Lane 1992), is observed to be a common, although possibly late event in tumour progression (Vogelstein *et al.* 1988). Amounts of cellular p53 have been observed to rise after DNA damage (reviewed by Lane 1993), and in mouse thymocytes from animals with a null p53 phenotype, created by homologous recombination, it was shown that DNA damage-induced apoptosis is a p53-dependent process, since the normal apoptotic response was lost in the absence of p53 (Lowe *et al.* 1993*b*; Clarke *et al.* 1993).

We have investigated the role of p53 in the apoptosis of stem cells in the crypts of the small intestine and colon (Merritt *et al.* 1994*a*). Immuno-histochemical analysis of the amount and cellular distribution of p53 protein in the crypts before and after DNA damage caused by radiation is shown in figure 6. The result is congruent with previous observations about the positional distribution of apoptotic cells: p53 protein was elevated in the stem cell region of the crypts of the small intestine whereas in the colon, the response to damage was more muted and the appearance of p53 protein after irradiation of the animals was not coincident with the position of the colonic stem cells. Most importantly, as figure 6 shows, the expression of p53 was coincident with the position of apoptotic cells. To determine the role of p53 in apoptosis in the intestinal crypts we have also utilized the null p53 animals (Donehower *et al.* 1992). Deletion of p53 completely prevented DNA

damage-induced cell death in intestinal crypts: in the small intestine of wild-type animals, the number of apoptotic events observed in 200 half crypts, 4.5 h after 8 Gy of irradiation, was 397. In the null p53 animals it was 10 (Merritt *et al.* 1994*a*). Essentially, the loss of p53 provided complete resistance to DNA damage-induced cell death and recent experiments have shown that the stem cell population of these irradiated p53 null animals is able to repopulate the crypt.

Consequently, in terms of the genesis, progression and therapy of colonic tumours our data provide a bleak picture, reflecting the clinical reality of this disease. Stem cells of the colon, likely to be the fount of such tumours, are characterized by their low levels of spontaneous and induced apoptosis because, first, they selectively express bcl-2 and secondly they have a poor DNA damage response in terms of the expression of p53. The high survival potential provided by this phenotypic background may promote carcinogenesis and progression, because of the tolerance to further DNA damage. The muted p53 response in normal colonic stem cells is then lost in more than 50% of colonic tumours (Vogelstein *et al.* 1988). Moreover, some of these tumours maintain expression of bcl-2 (J. A. Hickman, C. S. Potten, A. J. Merritt & T. C. Fisher, unpublished data). Expression of bcl-2 provides resistance to drugs such as fluorodeoxyuridine, as described above, despite the imposition of considerable DNA damage. Cells with DNA damage of the type introduced by fluorodeoxyuridine, but which have lost p53, do not die (Lowe *et al.* 1993*a*).

3. FUTURE DIRECTIONS

The minimal impact of chemotherapy on disseminated colon cancer, and on other solid tumours, is frustrating to the pharmacologist and clinician. Its impact on the cancer sufferer is obvious. How might the impasse be broken? The ideas of stimulus–response coupling and of survival thresholds, described above, may offer avenues for progress. That said, some caveats are important: first, the central tenet of chemotherapy is that it should be selective. Secondly, due recognition must be given, as stated above, to the problem of directing a therapeutic agent to a single locus in a multistage disease. With these goals and cautions in mind, it seems appropriate to consider strategies which modulate the survival thresholds of tumour cells. Naively, one might suppose that reducing the effects of *bcl*-2 for a short time might allow DNA damage induced by standard drugs like 5-fluorouracil to initiate cell death in colonic tumours expressing bcl-2. But, toxicity to other normal tissues, normally protected by bcl-2 expression, would presumably be superimposed. What is required is a colon tumour cell-specific perturbation under conditions of a reduced threshold for cell survival. The nature of the target – the 'stimulus' – might not be critical. Most pertinently, the observation that non-dividing cells, such as post-mitotic neurons, may be prompted to engage apoptosis permits that proliferation biochemistry need not be

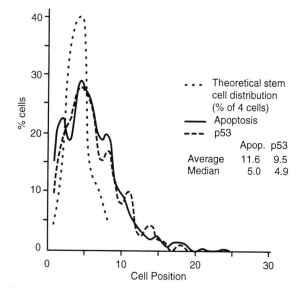

Figure 6. The relationship between the distribution of apoptotic fragments (solid line) and p53 positive nuclei (dashed line) along the length of the crypts of the murine small intestine 3 h after 8 Gy of irradiation. The theoretical positions of the stem cells, numbered from the base of the crypt, is also shown (dotted line). The average and median numbers of apoptotic cells was coincident with the expression of p53 protein.

a target. The idea of an acceptable target redundancy would also overcome concerns regarding the question as to which of the multiple changes associated with the progression of malignancy should be targeted.

However, it is the nature of these changes themselves which might be the tumour cell-specific trigger for apoptosis. The astonishing feature of many tumours is their tolerance of genetic abberations, observed as changes in ploidy, chromosomal deletions and translocations. Attenuation of the capability to tolerate and survive genetic abberations may itself initiate apoptosis. The attractive idea of using the multiple lesions of carcinogenesis as the selective trigger for tumour self-destruction may not be too long in the testing.

We thank our collaborators on the colon and drug resistance projects: Bill Moser, David Lane, Peter Hall, Chris Kemp, Allan Balmain, Gerard Evan, Caroline Dive, Chris Gregory, Anne Milner, Anne Jackman, Wynne Aherne and John Hartley. Support by grants from the Cancer Research Campaign is most gratefully acknowledged. J.A.H. also acknowledges support from Zeneca Pharmaceuticals.

REFERENCES

Castle, V.P., Heidelberger, K.P., Bromberg, J., Ou, X., Dole, M. & Nunez, G. 1994 Expression of the apoptosis-supressing protein *bcl-2*, in neuroblastoma is associated with unfavourable histology and N-*myc* amplification. *Am. J. Path.* **143**, 1543–1550.

Clarke, A.R., Purdie, C.A., Harrison, D.J., Morris, R.G., Bird, C.C., Hooper, M.I. & Wyllie, A.H. 1993 Thymocyte apoptosis induced by p53-dependent and independent pathways. *Nature, Lond.* **362**, 849–852.

Collins, M.K., Marvel, J., Malde, P. & Lopez-Rivas, A. 1992 Interleukin 3 protects murine bone marrow cells from apoptosis induced by DNA damaging agents. *J. exp. Med.* **176**, 1043–1051.

Colombel, M., Symmans, F., Gil, S., O'Toole, K.M., Chopin, D., Benson, M., Olsson, C.A., Korsmeyer, S.J. & Buttyan, R. 1993 Detection of the apoptosis-suppressing oncoprotein *bcl-2* in hormone-refractory human prostate cancers. *Am. J. Pathol.* **143**, 390–400.

Dive, C. & Wyllie, A.H. 1993 Apoptosis and cancer chemotherapy. In *Frontiers in pharmacology and therapeutics: cancer chemotherapy* (ed. J. A. Hickman & T. R. Tritton), pp. 21–56. Oxford: Blackwell Scientific Publications.

Dive, C. & Hickman, J.A. 1991 Drug target interactions: only the first step in the commitment to a programmed cell death? *Br. J. Cancer* **64**, 192–196.

Donehower, L.A., Harvey, M., Slagle, B.L., McArthur, M.J., Montgomery, C.A. Jr, Butel, S. & Bradley, A. 1992 Mice deficient for p53 are developmentally normal but susceptible to spontaneous tumours. *Nature, Lond.* **356**, 215–221.

Fisher, T.C., Milner, A.E., Gregory, C.D., Jackman, A.L., Aherne, G.W., Hartley, J.A., Dive, C. & Hickman, J.A. 1993 *bcl-2* Modulation of apoptosis induced by anticancer drugs: resistance to thymidylate stress is independent of classical resistance pathways. *Cancer Res.* **53**, 3321–3326.

Foulds, L. 1958 The natural history of cancer. *J. Chronic Dis.* **8**, 2–37.

Goligher, J.C. 1980 *Surgery of the anus, rectum and colon.* London: Balliere Tindall.

Hickman, J.A. 1992 Apoptosis induced by anticancer drugs. *Cancer Metast. Rev.* **11**, 121–139.

Hockenberry, D.M., Zutter, M., Hickey, M., Nahm, M. &

Korsmeyer, M.J. 1991 Bcl-2 protein is topographically restricted in tissues characterized by apoptotic cell death. *Proc. natn. Acad. Sci. U.S.A.* **88**, 6961–6965.

Ijiri, K. & Potten, C.S. 1987 Further studies on the response of intestinal crypt cells of different hierarchical status to eighteen different cytotoxic agents. *Br. J. Cancer* **55**, 113–123.

Ikegaki, N., Katsumata, M., Minna, J. & Tsujimoto, Y. 1994 Expression of bcl-2 in small cell lung carcinoma cells. *Cancer Res.* **54**, 6–8.

Kamesaki, S., Kamesaki, H., Jorgensen, T.J., Tanizawa, A., Pommier, Y. & Cossman, J. 1993 bcl-2 Protein inhibits etoposide-induced apoptosis through its effects on events subsequent to topoisomerase II-induced DNA strand breaks and their repair. *Cancer Res.* **53**, 4251–4256.

Lane, D.P. 1992 p53, guardian of the genome. *Nature, Lond.* **358**, 15–16.

Lane, D.P. 1993 A death in the life of p53. *Nature, Lond.* **362**, 786–787.

Leek, R.D., Kaklamanis, L., Pezella, F., Gatter, F. & Harris, A.H. 1994 Bcl-2 in normal human breast and carcinoma, association with ER positive EGFR negative tumours amnd in situ cancer. *Br. J. Cancer* **69**, 135–139.

Lowe, S.W., Ruley, H.E., Jacks, T. & Housman, D.E. 1993*a* p53-Dependent apoptosis modulates the cytotoxicity of anticancer agents. *Cell* **74**, 957–967.

Lowe, S.W., Schmitt, E.M., Smith, S.W., Osborne, B.A. & Jacks, T. 1993*b* p53 is required for radiation-induced apoptosis in mouse thymocytes. *Nature, Lond.* **362**, 847–849.

McDonnell, T.J., Troncoso, P., Brisbay, S.M., Logothetis, C., Chung, L.W.K., Hsieh, J.-T., Tu, S.-M. & Campbell, M.L. 1992 Expression of the protooncogene *bcl-2* in the prostate and its association with emergence of androgen-independent prostate cancer. *Cancer Res.* **52**, 6940–6944.

Merritt, A.J., Potten, C.S., Kemp, C.J., Hickman, J.A., Balmain, A., Lane, D.P. & Hall, P.A. 1994*a* The role of p53 in spontaneous and radiation-induced apoptosis in the gastrointestinal tract of normal and p53-deficient mice. *Cancer Res.* **54**, 614–617.

Miyashita, T. & Reed, J.C. 1993 Bcl-2 oncoprotein blocks chemotherapy-induced apoptosis in a human leukemia cell line. *Blood* **81**, 151–157.

Pezzella, F., Turley, H., Kuzu, I., Tungekar, M.F., Dunnhill, M.S., Pierce, C.B., Harris, A.L., Gatter, K.C. & Mason, D.Y. 1993 *bcl-2* Protein in non-small-cell lung carcinoma. *New Engl. J. Med.* **329**, 690–694.

Potten, C.S. 1992 The significance of spontaneous and induced apoptosis in the gastrointestinal tract of mice. *Cancer Metast. Rev.* **11**, 179–195.

Potten, C.S., Li, Q.Y., O'Connor, P.J. & Winton, D.J. 1992 A possible explanation for the differential cancer incidence in the intestine, based on distribution of the ctyotoxic effects of carcinogens in the murine large bowel. *Carcinogenesis, Lond.* **13**, 2305–2312.

Reed, J.C. 1994 Bcl-2 and the regulation of programmed cell death. *J. Cell Biol.* **124**, 1–6.

Twentyman, P.R. 1992 mdr-1 (P-glycoprotein) gene expression: implications for resistance modifier trials. *J. natn. Cancer Inst.* **84**, 1458–1460.

Vogelstein, B., Fearon, E.R., Hamilton, S.R., Kern, S.E Preisinger, A.C., Leppert, M., Nakamura, Y., White, R., Smits, A.M.M. & Bos, L. 1988 Genetic alterations during colorectal tumour development. *N. Engl. J. Med.* **319**, 525–532.

Walton, M.I., Whysong, D., O'Connor, P.M., Hockenberry, D., Korsmeyer, S.J. & Kohn, K.W. 1993 Constitutive expression of human *Bcl-2* modulates nitrogen mustard and camptothecin induced apoptosis. *Cancer Res.* **53**, 1853–1861.

15
Granulocyte apoptosis and the control of inflammation

C. HASLETT[1], J. S. SAVILL[2], M. K. B. WHYTE[3], M. STERN[3],
I. DRANSFIELD[1] AND L. C. MEAGHER[1]

[1]*Respiratory Medicine Unit, Department of Medicine (RIE), University of Edinburgh Royal Infirmary, Lauriston Place, Edinburgh EH3 9YW, U.K.*
[2]*Division of Renal & Inflammatory Disease, Department of Medicine, University Hospital, Nottingham NG7 2UH, U.K.*
[3]*Department of Respiratory Medicine, Royal Postgraduate Medical School, Hammersmith Hospital, Du Cane Road, London W12 0NN, U.K.*

SUMMARY

We have described a novel pathway available for the clearance of extravasated granulocytes from inflamed tissues whereby aging granulocytes undergo apoptosis, a process which leads to their phagocytosis by inflammatory macrophages. By contrast with necrosis, which may also be seen at inflamed sites, apoptosis represents a granulocyte fate which by a number of mechanisms would tend to limit inflammatory tissue injury and promote resolution rather than progression of inflammation: (i) apoptosis is responsible for macrophage recognition of senescent neutrophils with intact cell membranes which exclude vital dyes and retain their potentially histotoxic granule contents; (ii) the apoptotic neutrophil loses its ability to secrete granule enzymes on deliberate external stimulation; (iii) the macrophage possesses a huge phagocytic capacity for apoptotic neutrophils which it rapidly ingests and degrades without disgorging neutrophil contents; and (iv) the macrophage utilizes a novel phagocytic recognition mechanism which fails to trigger the release of pro-inflammatory macrophage mediators during the phagocytosis of apoptotic neutrophils. Preliminary characterization of the recognition mechanism implicates the integrin α v β_3 (vitronectin receptor) and CD36 (thrombospondin receptor) on the macrophage surface. Macrophage phagocytosis of apoptotic neutrophils is greatly influenced by the microenvironmental pH and by the presence of cationic molecules. Moreover, it can be specifically modulated by external cytokines and intracellular second messenger systems. By controlling the functional longevity of neutrophil and eosinophil granulocytes and their subsequent removal by macrophages, granulocyte apoptosis, with its potential for modulation by external mediators, is likely to play a key dynamic role in the control of the 'tissue load' of granulocytes at inflamed sites. Further elucidation of the mechanisms and control of apoptosis in granulocytes is likely to shed new light on the pathophysiology of inflammation and suggest new approaches to the therapy of inflammatory diseases.

1. INTRODUCTION: RESOLUTION VERSUS PERSISTENCE OF INFLAMMATION

It is now widely recognized that inflammation is central to the pathogenesis of a range of diseases which impose a heavy burden of mortality and untimely deaths in developed societies. In the lung these include chronic bronchitis and emphysema, asthma, and respiratory distress syndromes of the adult and neonate. Important inflammatory diseases in other organs include glomerulonephritis (the major cause of renal failure requiring dialysis), arthritides and inflammatory bowel disease. Most inflammatory diseases are characterized by the persistent accumulation of inflammatory cells which is associated with chronic tissue injury and scarring. In organs with delicate exchange membranes, such as the lung and kidney, these processes often result in catastrophic loss of function. However, it is also clear that, para-

doxically, inflammation has evolved as a highly effective component of the body's defences against infection and injury. Indeed, until the latter half of this century, inflammation was perceived as an entirely beneficial host response. Neutrophil and eosinophil granulocytes play essential protective roles in bacterial infections, such as lobar Streptococcal pneumonia, and parasitic infections, such as schistosomiasis. The acute inflammatory response in lobar Streptococcal pneumonia exemplifies the effectiveness of a rapidly mounted inflammatory response. In the pre-antibiotic era, Streptococcal pneumonia was widely prevalent, being responsible for more than 90% of pneumonias, yet the inflammatory response was effective enough to save the lives of more than 70% of patients. Perhaps more remarkable, given what we now know of the destructive and pro-fibrotic potential of neutrophils and activated macrophages, there was clear evidence that in more than 95% of cases lobar Streptococcal pneumonia resolved completely, with less than 2.5% progressing to fibrosis (Robertson & Uhley 1938).

It is reasonable to suppose that research aimed at determining how inflammation may normally resolve will not only provide important insights into the circumstances leading to the persistent inflammatory states which characterize most inflammatory diseases but will also suggest novel therapeutic strategies directed at promoting those mechanisms which favour resolution. By contrast with the initiation and amplification mechanisms of inflammation, however, comparatively scant attention has been paid to the processes responsible for its termination. Hurley (1983) considered that the acute inflammatory response might terminate by the development of: chronic inflammation; suppuration (abscess); scarring; or by resolution. Clearly, all the alternatives to resolution are potentially detrimental to organ function, but until very recently little research effort had been focused on the cellular and molecular mechanisms underlying the normal resolution processes of inflammation.

The resolution of inflammation is likely to be as complex as the initiation phase, but one pre-requisite is that extravasated inflammatory cells and their contents must be removed from tissues. We have been particularly interested in elucidating the mechanisms whereby granulocytes are cleared from inflamed sites.

2. THE TISSUE CLEARANCE OF EXTRAVASATED GRANULOCYTES

The neutrophil granulocyte is the archetypal acute inflammatory cell. It is essential for host defence, but it is also implicated in the pathogenesis of a wide variety of inflammatory diseases (Malech & Gallin 1988). It is usually the first cell to migrate to the scene of tissue perturbation, and a number of subsequent inflammatory events including monocyte emigration (Doherty *et al.* 1988) and generation of inflammatory oedema (Wedmore & Williams 1981) may depend on the initial tissue accumulation of neutrophils. Neutrophils contain a large number of agents with the capacity not only to injure tissues (Weiss 1989), but also to cleave matrix proteins into chemotactic factors (Vartio *et al.* 1981) with the potential to amplify inflammation by attracting more cells. Eosinophils play an important part in host defence against worms and other parasites, but they are also implicated in the pathogenesis of allergic diseases such as asthma. Although we have for some years been aware of the histotoxic potential of neutrophil and eosinophil contents there has been little formal study of the tissue fate of these cells. It is generally agreed that most extravasated neutrophils meet their fate at the inflamed site, but it had been widely assumed that they inevitably underwent disintegration (necrosis) before the fragments were removed by local macrophages (Hurley 1983). However, if this was the rule, healthy tissues would inevitably be exposed to large quantities of potentially injurious neutrophil

contents. Although a number of pathological descriptions have favoured neutrophil necrosis as a major mechanism operating in the inflammation, many of these examples have been taken from disease states rather than from 'beneficial' self-limited inflammation. Moreover, there has been evidence for over a century of an alternative fate for extravasated neutrophils, based on the original work of Metchnikoff, who, in vital preparations, was the first to describe the cellular events occurring during the evolution and resolution of the acute inflammatory response. Rather than neutrophil necrosis as the major mechanism during inflammatory resolution, he observed the ingestion of intact senescent neutrophils by macrophages (Metchnikoff 1891). Since then there have been several sporadic reports of macrophages phagocytosing neutrophils, and of particular relevance to the resolution of inflammation is the clinically described phenomenon of 'Reiter's cells': neutrophil-containing macrophages which have been described in synovial fluid from the inflamed joints of patients with Reiter's disease and other acute arthritides (Spriggs *et al.* 1978). In experimental peritonitis, macrophage ingestion of apparently intact neutrophils is clearly the dominant mode of neutrophil clearance (Chapes & Haskill 1983).

The mechanisms underlying these observations have only recently been addressed *in vitro*. Newman *et al.* (1982) showed that human neutrophils harvested from peripheral blood and 'aged' in culture were recognized and ingested by inflammatory macrophages (but not by monocytes) whereas freshly isolated neutrophils were not ingested. We have recently discovered that aging neutrophils and eosinophils (Savill *et al.* 1989a; Stern *et al.* 1992) constitutively undergo apoptosis and that this process is responsible for the recognition and ingestion of intact senescent granulocytes by macrophages.

3. APOPTOSIS IN AGING GRANULOCYTES LEADS TO THEIR PHAGOCYTOSIS BY MACROPHAGES

Neutrophils harvested from blood or from acutely inflamed human joints remain intact, retain their granule enzyme contents, and continue to exclude vital dyes for up to 24 h in culture. However, over this period there occurs a progressive increase in the proportion of cells exhibiting the classical light microscopical and ultrastructural (figure 1) features of apoptosis together with the 'ladder' pattern of chromatin cleavage which is indicative of endogenous endonuclease activation (Savill *et al.* 1989a). Only the apoptotic subpopulation of aged neutrophils is recognized and ingested by macrophages. Apoptotic neutrophils are not indestructible, and beyond 24 h in culture there is a progressive increase in the proportion of cells that fail to exclude vital dyes and spontaneous release of granule enzyme contents is observed. However, when neutrophils are cultured beyond 24 h in the presence of macrophages, the removal of apoptotic cells is so rapid and effective that no trypan blue positive cells are seen and there is no

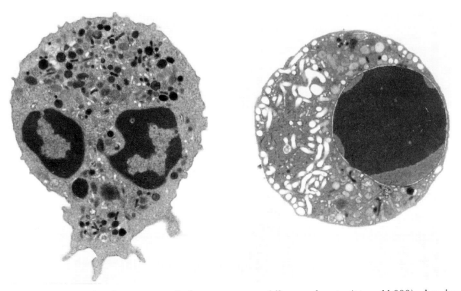

Figure 1. Electron micrograph of an apoptotic human neutrophil granulocyte ($\times ca$. 11 000) showing classical nuclear chromatin changes and dilatation of the endoplasmic reticulum (seen on the right) as compared with a non-apoptotic neutrophil (left).

release of granule enzyme markers into the surrounding medium (Kar *et al.* 1993). Macrophages *in vitro* can ingest and destroy several neutrophils with remarkable speed, such that in ultrastructural studies it is necessary to fix macrophages within minutes of the initial interaction between apoptotic cells and the macrophages in order to demonstrate recognizable neutrophils within phagosomes; thereafter ingested cells are no longer recognizable. This may represent part of the explanation why the dynamic contribution of this process to inflammatory tissue kinetics has not been fully appreciated until recently. Nevertheless, there are now several clear histological demonstrations of a role for apoptosis in the *in vivo* removal of granulocytes in acute inflammation. These include acute arthritis (Savill *et al.* 1989a), neonatal lung injury (Grigg *et al.* 1991) and experimental acute Streptococcal pneumonia during its resolution phase (figure 2).

Several lines of *in vitro* experimental evidence have emerged in support of the hypothesis that apoptosis provides a granulocyte clearance mechanism in tissue which would tend to limit inflammatory tissue injury and promote resolution rather than persistence of inflammation.

1. During apoptosis, the cell membrane remains functionally intact, as assessed by vital dye exclusion, and continues to retain cytosolic enzymic contents, but there is marked loss of a number of neutrophil functions, including secretion of granule enzymes after deliberate external neutrophil stimulation (figure 3). This suggests that the apoptotic neutrophil becomes 'functionally isolated' from external stimuli which would otherwise trigger responses with the potential to damage tissue (Whyte *et al.* 1993). We have recently shown that during neutro-

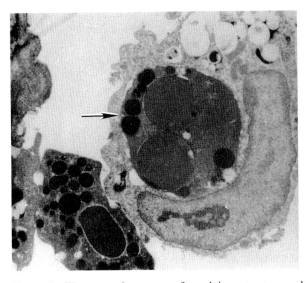

Figure 2. Electron microscopy of resolving streptococcal pneumonia showing a macrophage that has ingested an apoptotic neutrophil (arrow).

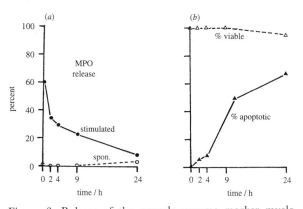

Figure 3. Release of the granule enzyme marker myeloperoxidase (MPO) during the *ex-vivo* culture of human neutrophils. Over 24 h there is a progressive increase in the proportion of apoptotic cells in the population but no significant necrosis (assessed by trypan blue exclusion) or spontaneous release of MPO. Moreover, with time the neutrophil population increasingly loses the ability to secrete MPO in response to deliberate external stimulation with the formylated peptide FMLP.

phil apoptosis there may be selective loss of surface receptors which exert important cellular functions (Dransfield *et al.* 1994). This down-regulation of neutrophil function would be expected to be particularly important if fully mature, competent phagocytes are not immediately available in the vicinity of neutrophils undergoing apoptosis.

2. Large numbers of apoptotic neutrophils can be cleared by macrophages without 'leakage' of potentially injurious neutrophil contents into the surrounding medium (Kar *et al.* 1993).

3. Although the usual response of macrophages to the ingestion of particles *in vitro* is to release mediators such as thromboxane, enzymes and pro-inflammatory cytokines, even maximal uptake of apoptotic neutrophils fails to stimulate the release of pro-inflammatory mediators (Meagher *et al.* 1992; figure 4). However, if apoptotic granulocytes are cultured beyond apoptosis to a point when they fail to exclude trypan blue, their ingestion by macrophages induces massive release of pro-inflammatory mediators (M. Stern, unpublished data). It was subsequently shown that this lack of a macrophage secretory response is not a function of the apoptotic body itself, but relates to the special mechanism by which the apoptotic cell is normally ingested (Meagher *et al.* 1992). These observations provided considerable impetus for our work on the molecular mechanisms responsible for macrophage recognition of apoptotic cells.

4. MECHANISMS WHEREBY MACROPHAGES RECOGNIZE APOPTOTIC NEUTROPHILS

In their early studies of the effects of sugars on macrophage recognition of apoptotic thymocytes, Duvall & Wyllie (1985) suggested that phagocytes possess a lectin mechanism capable of recognizing sugar residues on the apoptotic thymocyte surface exposed by loss of sialic acid. This mechanism does not appear to be involved in macrophage recognition of apoptotic granulocytes, but these findings stimulated our early work showing that recognition of apoptotic neutrophils was inhibited by cationic molecules, e.g. amino sugars, and was directly influenced by reductions of pH in a fashion implicating negatively charged moieties on the apoptotic neutrophil surface (Savill *et al.* 1989*a*). These observations were of considerable interest as a number of neutrophil and eosinophil products, e.g. elastase and major basic protein, are highly cationic and have been detected in tissues in inflammatory disease. Moreover, in sites of chronic inflammation, or in abscesses, tissue pH may be very low (Menkin 1956). Thus, local microenvironmental conditions at chronically inflamed sites could greatly retard macrophage removal of apoptotic neutrophils.

The observed amino sugar inhibition pattern led to a detailed series of investigations which implicated macrophage surface molecules, including the integrin $\alpha v \beta_3$ (the vitronectin receptor) (Savill *et al.* 1990) and CD36 (Savill *et al.* 1992) (a thrombospondin receptor), in the recognition and phagocytosis of apoptotic cells. These appear to link via thrombospondin, which acts as an intercellular bridging molecule, with an as yet uncharacterized recognition site on the apoptotic neutrophil surface. Recent studies by colleagues in Denver have been focused on changes in the apoptotic cell surface responsible for macrophage recognition. Their work suggests that macrophages may recognize phosphotidlyserine residues which become exposed on the surface of murine thymocytes induced by glucocorticoids to undergo apoptosis (Fadok *et al.* 1992). The *in vivo* significance of these observations is as yet uncertain, but it appears that the main difference between the two recognition systems relates to the utilization of alternative recognition mechanisms by different subpopulations of macrophages (Savill *et al.* 1993).

The definition of macrophage surface molecules responsible for apoptotic cell recognition suggests mechanisms by which neutrophil clearance may be regulated. Experiments so far have shown that a number of cytokines can promote macrophage uptake of apoptotic neutrophils (Ren & Savill 1993), and mediators which influence intracellular cyclic AMP may control the process through modulation of $\alpha v \beta_3$ function (McCutcheon *et al.* 1994).

5. REGULATION OF GRANULOCYTE APOPTOSIS BY EXTERNAL MEDIATORS

Histological observations in models of experimental Streptococcal pneumonia suggested that neutrophils

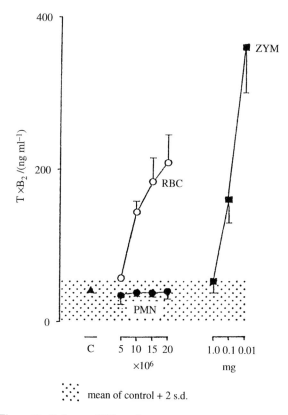

Figure 4. Release of Thromboxane B_2 from monolayers of human monocyte-derived macrophages after maximal phagocytosis of opsonized zymosan (ZYM), allogenic erythrocytes (RBC) or apoptotic human neutrophils (PMN).

at inflamed sites underwent apoptosis at a much slower rate than those derived from peripheral blood (C. Haslett, unpublished observations). This implied that factors present at the inflamed site might have retarded the inherent rate of neutrophil apoptosis. It has now been shown that the rate of neutrophil apoptosis *in vitro* is inhibited by a variety of inflammatory mediators including endotoxic lipopolysaccharide, C5a and GMCSF. Furthermore, inhibition of neutrophil apoptosis by these agents not only increased the lifespan of cultured neutrophils, but also greatly prolonged their functional longevity assessed by a number of parameters including chemotaxis and stimulated secretion (Lee *et al.* 1993). It had been known for several years that fibroblast-conditioned medium and specific growth factors, including GMCSF, could prolong the life of neutrophils and eosinophils in culture as assessed by failure of the cell to exclude the vital dye trypan blue (necrosis). Because we have clearly shown in healthy cultured granulocytes that apoptosis precedes ultimate necrosis of the cell, it seems likely that these previous observations can be explained by growth factor-induced modulation of the process of apoptosis. By modulating the lifespan and functional activity of neutrophils, apoptosis may represent a pivotal mechanism controlling their functional longevity at sites of inflammation. Experiments with eosinophils *in vitro* show that GMCSF inhibits eosinophil apoptosis, but that interleukin-5 is also extremely potent in this regard, whereas it has no effect on neutrophil longevity (Stern *et al.* 1992). It is intriguing that the apoptotic 'programme' should be under different controls in two such closely related cells.

There has been a great deal of recent interest in the role of intracellular signalling pathways and proto-oncogene expression in the control of apoptosis in a variety of cell types (e.g. Vaux *et al.* 1988). However, there has been comparatively little work on intracellular mechanisms controlling granulocyte apoptosis, and there are indications that internal controls in granulocytes may differ from those in lymphoid cells. In thymocytes, elevation of intracellular calcium by calcium ionophores induces apoptosis, and apoptosis induced by other stimuli, such as glucocorticoids is associated with rises in intracellular calcium (e.g. McConkey *et al.* 1989). However, in neutrophils spontaneously undergoing apoptosis there were no such rises in $[Ca^{2+}]i$ and agents increasing $[Ca^{2+}]i$ caused dramatic slowing of neutrophil apoptosis without inducing necrosis (Whyte *et al.* 1993). Furthermore, treatment of aging neutrophils with intracellular calcium chelators was associated with an increase in the rate of neutrophil apoptosis. Again in contrast to lymphoid cells, culture of neutrophils in the presence of inhibitors of protein synthesis, e.g. cycloheximide,

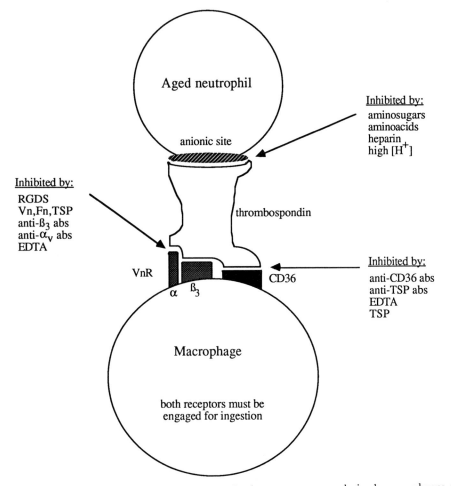

Figure 5. A model of the recognition mechanism whereby human monocyte-derived macrophages phagocytose apoptotic human neutrophils.

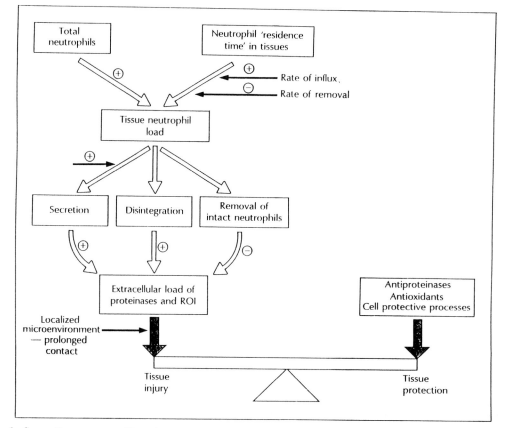

Figure 6. Some factors controlling the tissue 'load' of neutrophils at inflamed sites and the balance between injurious influences of inflammatory cells and tissue protective mechanisms.

caused an acceleration of the constitutative rate of apoptosis (Haslett *et al.* 1990). As yet there is little information on the role of proto-oncogenes in the control of neutrophil apoptosis.

6. HYPOTHESIS: A ROLE FOR GRANULOCYTE APOPTOSIS IN THE CONTROL OF INFLAMMATION?

The central role of apoptosis in the clearance of granulocytes from inflamed sites implies several levels of control but also the potential for disorders at various points of the clearance pathway which might promote inflammatory tissue injury and contribute to disease processes. Whether neutrophils meet their fate by disintegration, disgorgement of their contents, and phagocytosis by macrophages which respond by releasing inflammatory mediators (necrosis) or by removal of the intact senescent cell by macrophages which fail to release pro-inflammatory mediators (apoptosis) is likely to impinge on the precarious balance which normally exists between potentially damaging processes and tissue protective mechanisms in inflammation. While in all the spontaneously resolving examples of inflammation we have examined, the removal of whole granulocytes by apoptosis appears to be a major mechanism, examples of neutrophil necrosis are also seen. Therefore, it is possible that the balance between neutrophil apoptosis and necrosis at an inflamed site may represent a pivotal point in the control of tissue injury and in the propensity of an inflamed site to resolve or to progress.

By prolonging the functional longevity of neutrophils and eosinophils through inhibition of their constitutive rate of apoptosis, a variety of inflammatory mediators including growth factors and chemotactic cytokines exert important controls on the 'tissue load' of granulocytes (see figure 6). The removal of apoptotic granulocytes by macrophages is also under the control of inflammatory mediators including cytokines (Ren & Savill 1994) and agents which modulate macrophage cAMP levels (McCutcheon *et al.* 1994). Moreover, the phagocytic recognition mechanism is profoundly inhibited by physical conditions, including acidic pH and the presence of cationic molecules (Savill *et al.* 1989*b*) which may exist at chronically inflamed sites.

Finally, these observations may have some relevance for the development of novel approaches to anti-inflammatory therapy. With increasing knowledge of the internal mechanisms of apoptosis, it may be possible to specifically induce apoptosis in certain inflammatory cells at critical stages of the pathogenesis of inflammatory disease. The observation that apoptosis in such closely related cells as the neutrophil and eosinophil granulocyte appears to be controlled by different mechanisms lends some credence to this speculation.

REFERENCES

Chapes, S.K. & Haskill, S. 1983 Evidence for granulocyte-mediated macrophage activation after *C. parvum* immunization. *Cell. Immunol.* **75**, 367–377.

Doherty, D.E., Downey, G.P., Worthen, G.S., Haslett, C. & Henson, P.M. 1988 Monocyte retention and migration in pulmonary inflammation. *Lab. Invest.* **59**, 200–213.

Dransfield, I., Buckle, A.-M., McDowall, A., Savill, J.S., Haslett, C. & Hogg, N. 1994 Neutrophil apoptosis is associated with a reduction in CD16, FcγRIII (FcγR) expression. *J. Immunol.* (In the press.)

Duvall, E., Wyllie, A.H. & Morris, R.G. 1985 Macrophage recognition of cells undergoing programmed cell death. *Immunology* **56**, 351–358.

Fadok, V.A., Savill, J.S., Haslett, C. *et al.* 1992 Different populations of macrophages use either the vitronectin receptor of the phosphatidylserine receptor to recognize and remove apoptotic cells. *J. Immunol.* **149**, 4029–4035.

Grigg, J.M., Savill, J.S., Sarraf, C., Haslett, C. & Silverman, M. 1991 Neutrophil apoptosis and clearance from neonatal lungs. *Lancet* **338**, 720–722.

Haslett, C., Savill, J. & Meagher, L. 1990 Macrophage recognition of senescent granulocytes. *Biochem. Soc. Trans.* **18**, 225–227.

Hurley, J.V. 1983 Termination of acute inflammation. I. Resolution. In *Acute inflammation*, 2nd edn (ed. J. V. Hurley), pp. 109–117. London: Churchill Livingstone.

Kar, S., Ren, Y., Haslett, C. & Savill, J.S. 1993 Inhibition of macrophage phagocytosis in vitro of aged neutrophils undergoing apoptosis increases release of neutrophil content. *Clin. Sci.* **85**, 27P. (Abstract.)

Lee, A., Whyte, M.K.B. & Haslett, C. 1993 Prolongation of in vitro lifespan and functional longevity of neutrophils by inflammatory mediators acting through inhibition of apoptosis. *J. Leuk. Biol.* **54**, 283–288.

McConkey, D.J., Nicotera, P., Hartzell, P., Bellomo, G., Wyllie, A.H. & Orrenius, S. 1989 Glucocorticoids activate a suicide process in thymocytes through an elevation of cytosolic Ca^{2+} concentration. *Arch. Biochem. Biophys.* **269**, 365–370.

McCutcheon, J., Haslett, C. & Dransfield, I. 1994 Regulation of macrophage recognition of apoptotic cells by activation of protein kinases: redistribution of the integrin α v β$_3$. (Submitted.)

Malech, H.D. & Gallin, J.I. 1988 Neutrophils in human diseases. *N. Engl. Med. J.* **37**, 687–694.

Meagher, L.C., Savill, J.S., Baker, A., Fuller, R. & Haslett, C. 1992 Phagocytosis of apoptotic neutrophils does not induce macrophage release of thromboxane B$_2$. *J. Leuk. Biol.* **52**, 269–273.

Menkin, 1956 Biology of inflammation: chemical mediators and cellular injury. *Science, Wash.* **123**, 527–534.

Metchnikoff, E. 1891 Lectures on the comparative pathology of inflammation. Lecture VII. Delivered at the Pasteur Institute. Translated by F. A. Starling & E. H. Starling. New York: Dover, 1968.

Newman, S.L., Henson, J.E. & Henson, P.M. 1982 Phagocytosis of senescent neutrophils by human monocyte-derived macrophages and rabbit inflammatory macrophages. *J. exp. Med.* **156**, 430–442.

Ren, Y. & Savill, J.S. 1994 Mechanisms by which cytokines potentiate phagocytosis of neutrophils undergoing apoptosis. (Submitted.)

Robertson, O.H. & Uhley, C.G. 1938 Changes occurring in the macrophage system of the lungs in pneumococcus lobar pneumonia. *J. Clin. Invest.* **15**, 115–130.

Savill, J.S., Wyllie, A.H., Henson, J.E., Henson, P.M. & Haslett, C. 1989*a* Macrophage phagocytosis of aging neutrophils in inflammation. *J. Clin. Invest.* **83**, 865–875.

Savill, J.S., Henson, P.M. & Haslett, C. 1989*b* Phagocytosis of aged human neutrophils by macrophages is mediated by a novel "charge sensitive" recognition mechanism. *J. Clin. Invest.* **84**, 1518–1527.

Savill, J.S., Dransfield, I., Hogg, N. & Haslett, C. 1990 Macrophage recognition of 'senescent self'; the vitronectin receptor mediates phagocytosis of cells undergoing apoptosis. *Nature, Lond.* **342**, 170–173.

Savill, J.S., Hogg, N. & Haslett, C. 1992 Thrombospondin co-operates with CD36 and the vitronectin receptor in macrophage recognition of aged neutrophils. *J. Clin. Invest.* **90**, 1513–1529.

Savill, J.S., Fadok, V.A., Henson, P.M. & Haslett, C. 1993 Phagocyte recognition of cells undergoing apoptosis. *Immunol. Today.* **14**, 131–136.

Spriggs, R.S., Boddington, M.M. & Mowat, A.G. 1978 Joint fluid cytology in Reiter's syndrome. *Ann. Rheum. Dis.* **37**, 557–560.

Stern, M., Meagher, L., Savill, J. & Haslett, C. 1992 Apoptosis in human eosinophils. Programmed cell death in the eosinophil leads to phagocytosis by macrophages and is modulated by IL-5. *J. Immunol.* **148**, 3543–3549.

Vartio, T., Seppa, H. & Vaheri, A. 1981 Susceptibility of soluble and matrix fibronectin to degradation by tissue proteinases, mast cell chymase and cathepsin G. *J. biol. Chem.* **256**, 471–477.

Vaux, D.L., Cory, S. & Adams, J.M. 1988 *Bcl-2* gene promotes haemopoietic cell survival and co-operates with c-*myc* to immortalise pre-B cells. *Nature, Lond.* **335**, 440–442.

Wedmore, C.V. & Williams, T.J. 1989 Control of vascular permeability by polymorphonuclear leukocytes in inflammation. *Nature, Lond.* **289**, 646–650.

Weiss, S.J. 1989 Tissue destruction by neutrophils. *N. Engl. Med. J.* **320**, 365–376.

Whyte, M.K.B., Meagher, L.C., MacDermot, J. & Haslett, C. 1993 Impairment of function in aging neutrophils is associated with apoptosis. *J. Immunol.* **150**, 5123–5134.

Index